PRIVATE RIGHTS IN PUBLIC RESOURCES

Equity and Property Allocation in Market-Based Environmental Policy

Leigh Raymond

Resources for the Future
Washington, DC

An RFF Press book
Published by Resources for the Future
1616 P Street NW
Washington, DC 20036–1400
USA
www.rffpress.org

Library of Congress Cataloging-in-Publication Data

Raymond, Leigh Stafford.
Private rights in public resources : equity and property allocation in market-based environmental policy / Leigh Raymond.
p. cm.
Includes bibliographical references.
ISBN 1-891853-69-4 (hardcover : alk. paper) —
ISBN 1-891853-68-6 (pbk. : alk. paper)
1. Public lands—United States. 2. Concessions—United States. 3. Interest (Ownership rights)—United States. 4. Right of property—United States.
I. Title
KF5605.R39 2003
333.1'6—dc21

2003013942

f e d c b a

The paper in this book meets the guidelines for permanence and durability of the Committee on Production Guidelines for Book Longevity of the Council on Library Resources.

This book was typeset in *Giovanni* and *Myriad* by Carol Levie. The cover was designed by Marek Antoniak. Interior design by Naylor Design Inc.

About Resources for the Future *and* RFF Press

Resources for the Future (RFF) improves environmental and natural resource policymaking worldwide through independent social science research of the highest caliber. Founded in 1952, RFF pioneered the application of economics as a tool to develop more effective policy about the use and conservation of natural resources. Its scholars continue to employ social science methods to analyze critical issues concerning pollution control, energy policy, land and water use, hazardous waste, climate change, biodiversity, and the environmental challenges of developing countries.

RFF Press supports the mission of RFF by publishing book-length works that present a broad range of approaches to the study of natural resources and the environment. Its authors and editors include RFF staff, researchers from the larger academic and policy communities, and journalists. Audiences for RFF publications include all of the participants in the policymaking process — scholars, the media, advocacy groups, NGOs, professionals in business and government, and the general public.

CONTENTS

For my parents,
who first taught me about equity

PREFACE

When I entered graduate school in 1994, environmental policies relying on a so-called right to pollute were just gaining wider acceptance. The early achievements of the 1990 acid rain program, which created tradable rights to emit the pollutant sulfur dioxide, were making even former critics of these market-based policies begin to see their benefits. Formerly confined to a small band of economists and other policy scholars and advocates (many based at Resources for the Future), support for tradable permits and other market-based solutions to environmental problems seemed ready to hit the big time.

As a new student of environmental policy, I developed a growing appreciation of the market-based approach as I learned more about it. Something seemed to be missing, however, from the articles, lectures, and problem sets that constituted my initial introduction. Building an efficient and ecologically effective policy was a worthy goal, to be sure, but where were the issues of equity and distribution? How, in other words, would we decide who would get what share of the new environmental rights created by these market-based policies? Finding little work on this specific question in what was otherwise a rich and rewarding literature, I resolved to consider it in a scholarly manner. If no one else was studying this question, I declared with a certain degree of youthful hubris, then I would. The details of the task, however, soon brought me back to reality. Declaring an apparent gap in the literature is one thing, as seasoned scholars well-know, but finding a way to begin filling it is quite another. The critical challenge remained: how would I even begin to study the issue of equity in my own work?

Enter property theory. Early in my studies, I innocently began a term paper on a perennial question in natural resource policy: are federal grazing permits a form of private property? By the end of the paper, two things were clear to me: (a) that federal grazing permits were not legal property rights protected by the takings clause of the constitution and (b) that conclusion (a) didn't matter very much. In reality, it was clear that range users and others viewed their permits from the beginning as an entitlement, not a privilege, based in large part on decades of hard work on the land. More importantly, it was also clear that public policy had substantially respected those claims, for reasons that had little to do with takings doctrine and much to do with the influence of norms of ownership. As a budding policy scholar, it seemed to me that such norms of equity were a neglected aspect of the policy-making process—ideas that were far more important in the practice of politics than in the theory of it. It also seemed both interesting and important to understand which norms were influential under what circumstances and why. As Deborah Stone (2002, 34) says, "ideas are the very stuff of politics." So why not try, I decided, to describe more clearly what role ideas of fairness and equity play in shaping political outcomes, using property theory as the framework for analysis?

The result, to make a long story short, was a research project on market-based environmental policies that included ranch owners, utility managers, and seventeenth century property theorists among the *dramatis personae*. Eventually, climate change treaty negotiators joined the mix, bringing an international flavor to the proceedings. What stands out at the end of the process, however, is the common language of norms and property rights shared by these various actors, rather than the eclectic nature of the group. Equity norms, I conclude, play a critical role in politics, including in the crafting and implementing of market-based environmental policies. I hope that readers will find this book a worthy introduction to this political dialogue about equity, one that has been going on for centuries and seems likely to continue as long as governments create, and allocate, private rights in public resources.

Like most large projects, this effort would not have been possible without assistance from many sources. Much of the research was first completed at the University of California at Berkeley with the generous support of the U.S. Environmental Protection Agency's STAR fellowship program, as well as additional financial assistance from The Lincoln Institute of Land Policy and The Switzer Foundation. Substantial additional work on the manuscript was completed in Hyde Park, during a salutary two-year stint with the University of Chicago's best-kept secret, its Environmental Studies Program. Final revisions and additions were completed during my current employment at Purdue University. For all of this financial and institutional support, I am deeply grateful.

A work that uses historical information is likely to incur debts of gratitude to reference librarians, and this one is no exception. In particular, I thank Joan Howard at the National Archives in Denver, Richard Fusek at the National Archives in Washington D.C., Dave Hays at the University of Colorado at Boulder, Barbara Walton at the Denver Public Library's Western History Collection, and the staff at the American Heritage Center in Laramie, Wyoming for their assistance and guidance. I also benefited from the advice of Andy Senti and Jim Muhn at the Bureau of Land Management in Denver. Jim in particular gave me the benefit of his informed opinions and his extensive files on early public lands grazing issues. Edward Carpenter provided me many fine stories and unique sources of information regarding his father and the early days of the Division of Grazing during a stay in Grand Junction. I also have to thank Steve and Julia Brunner for their extended hospitality to a traveling researcher.

With respect to the acid rain program, I am indebted to the many individuals who agreed to be interviewed for this research: Phil Barnett, Dirk Forrister, Joseph Goffman, Judy Greenwald, Gary Hart, Ken Israels, Rusty Matthews, Brian McLean, John McManus, Sarah Wade, and Allan Zabel. Their unique views and insights added immeasurably to my understanding of this particular policy.

Others provided invaluable help in the writing of the manuscript. My original dissertation committee members John Dwyer, Louise Fortmann, and Kate O'Neill provided many hours of worthwhile feedback and inspiration. Pat Boling, Sarah Connick, Elizabeth Deakin, Charles Geisler, Lynn Huntsinger, Jonathan London, Sarah McCaffrey, Randal O'Toole, Jeff Romm, Lynn Scarlet, Jennifer Sokolove, Ted Steck, John Stella, Evalyn Tennant, Sissel Waage, Jeremy Waldron, and Felicia Wong all read and commented helpfully at various points in the writing process. Whatever its remaining imperfections, this book would be far the worse without their ideas and suggestions. Don Reisman at RFF Press has been extremely helpful and supportive ever since first approaching me at the 2001 APSA Annual Meeting. I am also grateful to Rick Freeman and the press's two other anonymous reviewers for their helpful comments on an earlier draft, as well as to Sally Atwater for her careful and skillful copyediting, as well as her good humor. Any mistakes or omissions that remain despite this capable assistance, of course, remain my own.

Three individuals merit special thanks. Laura Watt commented extensively on the entire manuscript more than once. Her enduring enthusiasm and sharp editorial eye sustained my efforts throughout the peaks and valleys of the initial writing process. My dissertation advisor, Sally K. Fairfax, also made bountiful comments on more than one version of earlier drafts. I am deeply grateful for her high standards, her collaborative approach, and her unflagging encouragement on this and other projects. Finally, Teresa Raymond provided support ranging from research assistance for a scholar with a broken hand to detailed comments on the entire manuscript. For her many contributions, to this effort and otherwise, I am eternally thankful.

Leigh Raymond

INTRODUCTION

If we are to have any hope of pursuing equitable cooperation, we must try to arrive at a consensus about what equity means.

— *Henry Shue, Global Environment and International Equity*

Environmental issues are distributive issues. For a member of Congress, an air pollution bill allocates health risks and economic development across the nation. For a Forest Service administrator, a forest management regulation decides who gets access to public lands on what terms. For a multilateral treaty negotiator, the wording of an international climate change protocol determines which nations bear the greater burden of reducing greenhouse gas emissions. For a local activist, a protest against siting a hazardous waste incinerator is a statement about who bears the impact of toxic pollution and waste. By allocating environmental benefits and burdens, all of these decisions raise important issues of equity, or "distributive justice" in philosophical terms. Some policies and decisions confront these equity implications head-on, whereas others ignore or avoid them, but they all have similar distributive implications.

Examples in which equity ideas play a prominent public role in environmental policy are numerous. The modern environmental justice movement protests the unequal distribution of environmental hazards among races and socioeconomic classes (Bullard 1990). Transboundary pollution issues, such as climate change and acid rain, have caused high-profile equity-based conflicts between regions and nations (Albin 1995; Rose 1990). Restrictions on private property owners to preserve endangered species habitat or limit urban sprawl are also frequently challenged or defended on the basis of distributive justice (Epstein 1998; Sax 1983). As resource scarcity increases, it will be even more difficult to approach any environmental policy without addressing such equity implications directly; no longer can they be ignored or obscured.

The growing prominence of equity issues in environmental policy inspires this book, which asks, how do politicians and other political actors respond to these concerns? How, in other words, do equity principles influence political behavior in creating and implementing environmental policy? At the outset, several possible answers present themselves. Politicians and agency staff could downplay and deemphasize the equity implications of their decisions. They could pursue their own political agendas without reference to equity or fairness at all. They could use the language of equity and fairness as a smokescreen for their own self-interested goals. Or, in what seems like a potentially naïve answer, they could at least partially shape their choices based on certain well-accepted principles of fairness and distributive justice.

At the risk of being labeled a Pollyanna, in this book I make a case for the naïve alternative. In some situations, I argue, specific equity ideas have a powerful effect on the political process. These equity ideas can be referred to more formally as norms of fairness. In general, a *norm* is a nonlegal rule of behavior that is culturally determined, commonly held, and socially enforced (Coleman 1990, 250; Ellickson 1991, 127–30). Norms can govern a wide range of behaviors, from clothing choices ("Don't wear jeans to church") to conversation topics ("Don't ask others about their personal finances") to gender roles ("Ladies first"), but many focus on actions required by principles of fairness. The Golden Rule—Do unto others as you would have them do unto you—is a good example of a widely held ethical norm.

Despite my focus on normative ideas, it is critical to note that my task in this work is empirical rather than normative. Other authors (e.g., Sen 1992;

Sagoff 1988; Kelman 1981) have made strong arguments for why equity considerations should matter in economics and politics. They generally argue *deductively* and *normatively* in favor of the explicit inclusion of principles of equity in public policy. This project takes a different approach and seeks instead to understand *inductively* and *empirically* how equity actually enters the public policy mix. The chapters that follow support the proequity literature not by showing that equity ideas *should* influence public policy, but by showing how in fact they already do.

This is not to argue that the present undertaking lacks normative inspiration or implications: encouraging and facilitating the explicit inclusion of equity ideas in public policy is an acknowledged purpose of this work. As an empirical study presenting a model of how equity affects policy, however, it complements those more abstract efforts as well as studies focused on efficiency as a policy goal. Policymakers struggle with equity questions all the time, and normative theory can help us understand and describe those struggles in greater detail. Ideally, the results of this study will help stimulate and inform arguments from both the traditional economics perspective and the more philosophical approaches to considering models of political behavior.

Perhaps surprisingly, the policy situations I consider in this book are ones in which equity is less frequently mentioned by scholars: the creation of market-based environmental policies based on economic incentives. Market-based environmental policies gained popularity in the 1980s and 1990s, primarily in a series of U.S. regulatory experiments in air pollution control culminating with the acid rain program (Title IV) of the Clean Air Act Amendments of 1990. The acid rain policy created tradable private rights, or "allowances," to emit the pollutant sulfur dioxide (SO_2). Although they include many traditional powers of private ownership, SO_2 allowances are like mere legal licenses in that they remain subject to cancellation by the government at any time without compensation. In this respect, they represent a type of quasi-property right that I will refer to as *licensed property.*

Heralded as unprecedented at the time, the acid rain program actually resembled previous government policies creating similarly "licensed" private property rights in public resources. One such historical case is the allocation of federal grazing permits under the Taylor Grazing Act of 1934. This act created a new agency, the Division of Grazing, to bring federal control over private livestock on the public domain for the first time. The division's task was to end overgrazing and often violent conflicts over public domain forage by requiring federal grazing permits for all users. Grazing permits, like SO_2 allowances, were an attempt to create something like private property rights in a public resource while reserving the power of government to revoke or adjust those rights in the future. They were, in other words, another example of licensed property.

Nor are allocations of licensed property limited to the past. Inspired by the apparent success of the acid rain program, international negotiators seeking to limit emissions of so-called greenhouse gases considered responsible for climate change have also turned to the licensed property model. The continuing negotiations have resulted in one significant allocation so far under the terms of the 1997 Kyoto Protocol. The distribution of these rights, finalized in a late-night negotiating session in Kyoto, was actually the outcome of a series of meetings and discussions among more than 100 nations lasting two and a half years. The allocation of emissions rights to greenhouse gases represents a modern, and ongoing, attempt to create private property–like rights in a public resource. The climate change case will thus form a second point of comparison with the acid rain program.

A critical policy question in all three cases was the initial allocation of rights: legislators, administrators, and negotiators had to determine what share of the public resource should be granted to each potential user. As a focus for distributional concerns, the initial allocation of licensed property is the specific subject of this book. Interestingly, these allocations are sometimes neglected in treatments of the acid rain program and other market-based policies. The oversight is partly due to the widespread influence of economist Ronald Coase's 1960 article "The Problem of Social Cost," which argued that the initial distribution of property rights was irrelevant to reaching an economically efficient outcome (given certain assumptions about information and transaction costs). Since economists publish much of the research on market-based policies like the acid rain program, the subsequent lack of attention to the initial allocation is not surprising. Although there has been more interest of late in theoretical discussions of the initial allocation, there remains far less attention to how these allocations unfold in practice (Cramton and Kerr 1998; ELI 1997). Those researchers who do consider actual allocation experiences frequently criticize them as distributions that "grandfather" rights based on current levels of resource use, or otherwise interpret the process as an example of rent-seeking and undue political control by private interests.

For those interested in the policymaking process, however, understanding the initial allocation in these cases is crucial—it is the issue that generally makes or breaks the political adoption of any licensed property policy. And understanding the initial allocation process, I will argue, requires one to study the role of norms of fairness. Central to this analysis is defining the content of the relevant norms. Those that influence political behavior in these cases are much more specific than the Golden Rule. Policies creating and allocating licensed property rely on detailed norms governing the right to private property. Philosophers and political theorists have struggled for centuries to formulate normative principles and justifications for claims of private ownership, and the literature provides a fertile source of private property concepts that remain influential.

Thus, this book seeks to link political philosophy and explanations of political behavior in the manner suggested by several public choice theorists interested in the influence of norms on political action (Noll and Weingast 1991; Almond 1991; Elster 1990). It argues not only that specific norms of ownership effectively guided political behavior in the three cases considered here, but also that in specific contexts, future initial allocations of licensed property will generally follow similar norms. In other words, this book is not simply a *post hoc* account of the cases discussed; it aims to describe how allocations are likely to proceed in future situations as well. By proposing a more general model of norms at work, this work aspires to follow in the tradition of Ellickson (1991) and other scholars seeking to document the role of norms in human behavior in a more theoretical and less anecdotal manner.

The data used to support the analysis in the three cases are written and oral accounts of the administrative, legislative, and diplomatic wrangling over the allocation issue. Although the actual allocation rules are important, they are just the proverbial tip of the iceberg in providing useful information for studying the influence of equity principles. Administrative correspondence, lawsuits, public hearings, and private meetings all provide insight into the process by which the final set of distributive principles and outcomes was created. The written and oral records of these steps toward the final rules serve as crucial sources of information.

The research argument has three distinct components. First, data from the sources just described document the *rhetoric of equity* common to all three allocation debates. In particular, the research shows that in all three cases, forms of licensed property were created to address ecological problems, and that political actors raised numerous equity arguments regarding the distribution of those rights. Of course, the prevalence of appeals to fairness is inconclusive evidence of the actual policy influence of these ideas. The second component of the argument goes beyond the rhetoric to demonstrate how the final *legal rules of allocation* in each case carefully followed a few specific equity norms. This second step connects the rhetoric of equity to actual distributions of rights.

The third component of the argument addresses the content of the norms in evidence. Surprisingly, it suggests that the same *framework of property norms* shaped the allocation process in all three cases. In the two U.S. cases considered, this similarity is especially noteworthy. The study documents how similar the rhetoric was in the acid rain and grazing cases, gravitating toward the same normative positions. It also shows how a similar process over time led to the final set of allocation rules in each case. Both sets of decisionmakers started with rules based on an "intrinsic" theory of property grounded in the ideas of John Locke, and then moved toward an opposing "instrumental" view (in the spirit of legal philosopher Morris Cohen) without abandoning the intrinsic position. Eventually, the allocation outcome in each U.S. case represented an attempt to synthesize these two contradictory norms of ownership. In neither case does the final allocation resemble grandfathering of the status quo, as has been reported. Instead, it combines conflicting intrinsic and instrumental principles in a process evocative of the nineteenth-century German philosopher G. W. Hegel's account of ownership.

In the climate change example, a different conflict unfolded within the same property theory framework. Here, the conflict was between two alternative views of property: one based on simple possession, as promulgated by the eighteenth-century Scottish thinker David Hume, and one based on a radical egalitarian perspective, advocated by the French theorist P.J. Proudhon. The differences in the allocation arguments in the climate change case seem to derive from contextual factors, including national disparities in dominant property and equity norms, global inequalities of wealth, and the nature of the prior resource use at issue. In this ongoing allocation, no reconciliation between opposing principles has yet been achieved.

At this point readers may wonder at the image of bureaucrats, members of Congress, or treaty negotiators debating the property theories of eighteenth- and nineteenth-century philosophers. And of course, no such discussion has occurred or is likely in the future. My contention is not that political actors directly invoked political theorists like Locke and Hegel in making their allocation arguments and decisions. Instead, the arguments and outcomes of these processes echo the ideas of certain prominent theorists and reflect the substantial influence these writers have had on the world's political and legal culture. Political actors do not create relevant norms out of whole cloth; they work within established normative frameworks that both reflect and have been shaped by influential political theorists. Hegel and his brethren make no direct appearances in the political record of these cases, at least none that I am aware of, but their ideas effectively summarize the various positions represented.

In appealing to Hegel for some guidance, I am inspired by and rely heavily on the work of others who have used his ideas to address similar kinds of apparent contradictions in law and politics. Both Margaret Jane Radin (1982) and Alan Brudner (1995), for example, have used Hegel's theory of ownership to explain certain puzzling or confusing trends or ideas in property law. Brudner in particular sheds light on the Hegelian process of defining property rights through a reconciliation of opposing principles. Jane Bennett (1987) has sought similar guidance from Hegel to understand and break the deadlocks in modern environmental policy. In all three cases, the Hegelian emphasis on a dialectical understanding of opposing ideas helps unify apparent contradictions that threaten to stalemate a legal or policymaking process. Bennett creates the wonderful image of a "fractious holism" based on Hegelian thought—a unity of opposing ideas that is far from harmonious or easy to achieve. I propose that policymakers actually seek such a fractious holism of ideas to create and distribute licensed property in a manner that respects both intrinsic and instrumental norms of allocation.

The relevance of the property theory framework to all three distributions is especially remarkable given the different historical and ecological settings involved. The Taylor Grazing Act allocation, for instance, took place during the Great Depression under very different political and economic conditions from those surrounding the acid rain bill nearly 60 years later. The relevant private interests in each case—livestock owners and electrical utilities—have little in common, the ecological issues of overgrazing and air pollution are distinct, and the battle over allocation principles was primarily administrative in the grazing case but legislative in the acid rain program. Yet despite these differences, the same norms seem to have played a central role in both cases.

Moving outside the United States, the contextual differences are even more dramatic. In the climate change setting, the actors are nation-states rather than individuals, and the conflict involves representatives from more than 100 nations with distinctive normative traditions and ideas. The ecological issue—the overuse of a global resource to the detriment of the populations of all nations—is unprecedented in complexity and scale. And the allocation process takes place through diplomacy rather than administrative action or legislation. Yet again, the property norms framework describes the allocation arguments and process quite well, albeit in terms of a different set of opposing theories than were prominent in the U.S. examples.

Of the many contextual factors that shape an allocation, this study argues, differences in the kind of prior resource use are particularly relevant. Grazing, for example, is a more tangible form of resource use than air pollution: it is more visible and spatially concentrated, qualities that appear to have important implications for the allocation process. Grazing is also more easily perceived as a beneficial use than air pollution: cowboys are symbolic heroes and can draw on social approval much more than utility executives and power plant operators. I hypothesize that the more people perceive the resource use as tangible and beneficial, the more allocations will favor claims based on prior use. Instrumental arguments, by contrast, will be relatively more successful for uses that are further from the Lockean intrinsic ideal. If the prior use is widely perceived as without benefit, as in the climate change example, still other norms are likely to come into play, including Humean and more radical egalitarian options.

The vital role of property norms in these cases has wider implications for students and practitioners of public policy. Two are conceptual. First, policy instruments like grazing permits and emissions allowances have rarely been connected in the literature of public policy, and yet they share crucial qualities. The idea of licensed property proposed here helps link these policies conceptually and facilitates the application of the property theory framework to similar cases in the future. Second, the prominent role of equity norms in these cases indicates that they are an important, and understudied, determinant of political behavior. The research therefore supports the larger idea that political scientists and policy scholars would do well to pay greater attention to the role of normative ideas in shaping political actions and outcomes.

The third implication is more practical: the substantial political influence of specific equity norms suggests that the range of possible distributions of use rights may be much narrower in practice than in theory. In the abstract, one can separate the distributive question from the issue of choosing a market-based public policy, since any initial allocation is possible (and equally efficient under the terms of the Coase Theorem). In practice, the range of possible allocations is substantially limited by certain powerful equity norms. For example, a distribution of rights starting from Lockean principles and following the Hegelian process is likely to occur under a policy in which U.S. decisionmakers play a controlling or dominant role. Or, to take another example, an auction is unlikely to succeed on equity terms in an allocation setting where significant amounts of "beneficial" prior use have occurred. This outcome means that choosing a market-based policy based on licensed property favors certain distributive claims on the resource and weakens others. The policy instrument, in other words, implies a much narrower range of allocation outcomes in practice.

This book makes its case for the importance of equity norms as follows. Chapter 1 defines such terms as *market-based policy* and *licensed property* in more depth and suggests the study's broader implications by locating the grazing, clean air, and climate change cases within a larger group of licensed property examples. It also briefly reviews the treatment of norms and initial allocations in the existing literature on public choice and market-based policies. Chapter 2 gives detailed content to the norms considered in this study, building a framework of property theories including the intrinsic, instrumental, Humean, egalitarian, and Hegelian approaches. It also reviews a variety of possible allocation strategies for licensed property policies and places them within this framework.

The book then applies the property framework to the empirical data. Chapter 3 describes the process of allocating licensed property under the Clean Air Act Amendments of 1990. Chapter 4 presents the parallel allocation process almost 60 years earlier under the Taylor Grazing Act. Both chapters explore the significant public and private effort devoted to allocation questions on equity grounds and demonstrate how a similar allocation process and outcome best described as Hegelian occurred in both cases. Chapter 5 then applies the property framework in an international context—the allocation of emissions rights as part of the ongoing negotiations over the problem of climate change. Finally, Chapter 6 summarizes the ability of the framework to explain the empirical cases, as well as the broader policy implications of the results.

A final point about structure. Although I have written the book on the assumption that readers will enjoy each chapter in the order presented, the text covers a diverse range of topics. Those interested in a particular empirical case may be tempted to skip ahead and read one of the later chapters separately from the rest of the book. As a general rule, this should be possible without too much difficulty: each empirical chapter can stand more or less on its own, with one important exception. Given that the book's theoretical framework is developed in Chapter 2, readers who skip that discussion may find some of the property theory terminology used in Chapters 3, 4, and 5 unclear or puzzling at times. At least a brief review of the second chapter, therefore, will significantly aid understanding of any of the three subsequent empirical cases and is strongly encouraged even for those primarily interested in a specific discussion of climate change, public lands grazing, or the acid rain program.

CHAPTER 1

THE POLITICS OF LICENSED PROPERTY

When we talk about fairness ... my observation in the Congress is that fairness is one more vote, and that them that has, gets; and that it helps to have the gavel to do the getting.

—*Representative Joe Barton, Texas*

Equity is not an especially popular topic in modern, empirical research on public policy and the political process. Indeed, many politicians and scholars of public governance seem to agree with Representative Barton's "gavel theory" of fairness in politics, presented in the epigraph above. Equity, from this perspective, is simply a smokescreen for the naked exercise of political power in pursuit of self-interested ends. Pork-barrel politics—the careful tending to the economic needs of one's constituents—is the name of the game.

Few would deny that there is an important element of truth to that description of the legislative process. Yet the literature on political action seems unsatisfied with explanations of political behavior based exclusively on this model. Indeed, theories of public choice are now increasingly inclined to acknowledge the role of norms, including norms of fairness, in shaping political behavior. This chapter considers the current treatment of equity and norms in literatures on public choice and market-based policies. It offers some possible explanations for equity's limited (although currently expanding) role in these works at present, as well as a further justification for why that limited role is an important oversight. In this manner, the chapter explains how the present study fills a gap in the empirical research on market-based policies while building on promising leads in the more theoretical literature of public choice.

The chapter proceeds in five sections. First it presents more detailed definitions of important terms, such as *market-based policy* and *licensed property.* Next the chapter delves briefly into property theory to defend licensed property rights as a legitimate form of private ownership. Third is a discussion of case selection that demonstrates the study's broader policy implications by locating the Taylor Grazing Act, Clean Air Act Amendments, and climate change cases within a larger group of policies involving licensed property. Fourth, the chapter considers attention (or inattention) to equity in existing work on rational choice theory, market-based policies, and these specific cases to highlight the important theoretical ideas and empirical gaps that inspire this book. Finally, the chapter considers and rejects a pair of common objections to studying the role of equity in public policy.

MARKET-BASED POLICIES AND "LICENSED PROPERTY"

Uncontrolled use of a natural resource is a common environmental problem. Initially, many resource users—be they livestock owners, fishermen, or polluting industries—operate during a period of effective resource abundance, in which regulation is unnecessary. When increased usage leads to significant environmental degradation, however, the need for some form of control becomes apparent. Under certain conditions, such control arises from nongovernmental mechanisms relying on norms of behavior and mutual monitoring and enforcement (Ellickson 1993). Contrary to the dire predictions of Garret Hardin's (1968) "tragedy of the commons" paradigm, these nonlegal forms of control flourish in various settings worldwide (Agrawal 2002; Ostrom 1990; McCay and Acheson 1987). Such "common property" regimes remain an important part of domestic and international resource management policy.

Despite the extensive success of common property management, in the absence of certain factors effective private control over resource use can fail to materialize. In such circumstances, government regulations become an important policy choice. Government can control use of a natural resource in several ways. Most frequently, government imposes specific and inflexible legal restrictions on time or degree of use, relying on executive branch action for enforcement. Command-and-control regulations of this type may, for example, impose limits on season of use, restrictions on the type of use permitted, or limits on the amount of resource consumption per user. Familiar examples of the command-and-control approach include rules that limit fishing and hunting to specific seasons, prohibit doe hunting, prohibit grazing by a specific type of livestock, or require installation of a specific technology to limit air pollution.

An alternative form of government regulation relies more on market principles. Rather than controlling every user through a system of rules regarding place, time, and degree of utilization of the resource, these market-based policies provide new economic incentives and disincentives to the user. In this study, a *market-based policy* is accordingly defined as any policy that relies primarily on new economic incentives or disincentives created by governments to control resource use. Such policies aim to be more flexible and economically efficient than their command-and-control brethren. Command-and-control tells the user exactly how to solve the problem: the fisherman can catch flounder only two weeks a year and the power plant must install a flue scrubber that captures 90% of the fly ash generated. A market-based approach provides a revised economic framework within which users calculate the best way of meeting the environmental goal: the fisherman can harvest his share of the flounder catch each year whenever and however he wants, and the power plant operator is free to reduce emissions below the required level and then sell her surplus emissions rights to a utility with dirtier equipment. In this manner, market-based approaches are intended to lower the total social costs of a given level of environmental improvement.

Specific market-based policies include pollution charges or taxes, deposit refund systems (such as state "bottle bills"), and tradable permits. An emissions tax creates an economic disincentive that raises a factory owner's cost of polluting the air to a more socially optimal level. A deposit refund system discourages undesired resource use (such as burying recyclable materials in a landfill) by providing an incentive (the returnable deposit) that encourages alternative actions. A permit for a fixed portion of the total fish catch (or of the total forage on the federal range) tries to encourage more responsible resource use as well, by strengthening the user's private rights over the resource. Users with stronger private rights stand to reap the benefits of their ecologically responsible behavior in future years, rather than lose out to other competitors. In each case, the policy aims to internalize for the user the additional costs of resource degradation.[1] In this manner, such policies aim to encourage the appropriate degree of environmentally sensitive behavior.

Tradable permits differ from other market-based approaches in that they create individual legal rights to a resource, thereby restricting access by others. The previously mentioned flounder fisherman, for example, effectively owns his share of the total allowable catch. These permits are akin to private property, frequently including many of the traditional rights of ownership, with one crucial exception. In nearly every case, they are subject to future cancellation or modification by the government without compensation to the owner. This ongoing vulnerability makes it misleading to simply call such rights private property, which in the United States would be protected against public seizure under the takings clause of the Fifth Amendment to the U.S. Constitution. Yet the rights in question are otherwise very much like private property. Much more than a bare license—a weak legal right lacking any property-like qualities—they are a unique form of ownership that exists at the discretion of public officials. They are property, in other words, that has only been licensed to private owners rather than given or sold to them as a fully vested legal right. They are, in short, "licensed property."

More formally, this work will use the term *licensed property* to mean a private legal right that provides a significant degree of security and exclusivity to resource users but remains unprotected from future government adjustment or cancellation without compensation. Of course, licensed property is not a term found in any legal text on property law, nor is it being proposed as such. Instead, it is suggested here as a concept for political analysis, where the idea of a licensed form of private property would be quite useful. The term recognizes that the private rights created by certain market-based policies are intended to function as a form of private ownership. By calling such rights a form of property, the ideas of property theory can be brought to bear on the subject, as will be done in this work. Yet the adjective *licensed* takes the question of government compensation off the table.

Thus, the notion of licensed property permits one to identify these rights in political discourse as a form of private ownership without implying that compensation is required, or that every possible power of property is included in the term. This work will show how those debating licensed property policies frequently have asked the question, "Is this a property right?" Given the variety of possible powers of ownership, a better question would be, "What kind of property right is this?" The explicit flexibility of the phrase *licensed property* would help steer policy discussions in this more fruitful direction.

Policies creating licensed property rights are clearly market based; they rely on the new incentives created by private ownership to attain environmental improvements. The specific incentives for the licensed property owner vary, however, with the type of right created. Licensed property rights may or may not include other traditional powers of ownership, including transfer, bequest, and destruction. Tradability, for example, adds one important incentive for users to make use reductions where they are least costly and to move permits where they are valued most highly. Of course, trading is only one way to provide incentives to owners of these rights. Indeed, not all licensed property policies encourage trading to the same degree. In fact, some limit trading so severely, either intentionally or unintentionally, that they prevent any exchanges from occurring (Hahn 1989). Even in the leading case of the U.S. SO_2 program, trading has had a relatively limited impact: many utilities are complying with the law without trading allowances at all (Burtraw 1998; Burtraw and Swift 1996).

Regardless of their degree of liquidity, however, licensed property policies provide certain other advantages of private ownership to their holders. All licensed property policies rely on *security* and *exclusion*, for example, in their attempts to achieve environmental improvement and gain regulatory efficiency. Consider instruments like individual transferable quotas for fisheries or Bureau of Land Management grazing permits for federal range users. Quotas, as in the flounder example above, discourage the inefficient and ecologically harmful "race for the fish"[2] by creating a secure licensed property right for each user to a percentage of the total allowable catch. Similarly, grazing permits encourage more efficient and responsible resource use by granting security to the range user, preventing an ecologically destructive "race for the grass." The power of transfer is irrelevant to these particular incentives, a point that is sometimes overlooked.[3]

IS LICENSED PROPERTY REALLY PROPERTY?

Can one really describe items like grazing permits and emissions allowances as forms of private ownership? The Clean Air Act Amendments of 1990 include an explicit statement that allowances are not a form of private property, and the courts have ruled conclusively that federal grazing permits are not "vested rights" of private ownership (Raymond 1997). Many of these licensed property rights lack the full complement of the traditional capitalist powers of ownership, including unrestricted transfer or perpetuity of tenure. How, then, in light of these legal pronouncements and facts, can one reasonably discuss them as a form of property?

Legally, there is no question that licensed property rights are not private property—at least not in the sense defined by the Bill of Rights. The takings clause in the U.S. Constitution protects vested rights of ownership from government confiscation without compensation. Licensed property is often explicitly disqualified from compensation under this clause; the government reserves the power to revoke these rights without payment at any time.[4] In the international context, there is even less support for the idea that a licensed property right created by an international treaty would somehow be the legal equivalent of private property, as is discussed in Chapter 5. Nevertheless, despite the threat of revocation, licensed property rights often exhibit remarkable staying power. The *de facto* permanence of many of these rights is an important counterpoint to their insecure *de jure* position.

The Western jurisprudence of property law provides further argument against a taxonomy of ownership that includes licensed property. The American definition of private property is rooted in a legal tradition dating back to English common law. Drawing on the ideas of Blackstone and others, this Anglo-American idea of ownership stresses the nearly unbounded rights of the owner against all others (Dwyer and Menell 1998, 273). Several aspects of ownership are regarded as central: property must be fully exclusive, transferable, and secure against confiscation by both government and other private individuals. Rights of private ownership with these three qualities are the ideal type for most economists seeking greater social efficiency (e.g., Tietenberg 1988, 39).

Despite this tradition, the realities of modern property law are a far cry from the Blackstonian ideal. Property is a highly divisible right. Since the late nineteenth century, jurists and scholars have frequently portrayed property as a bundle of rights, or "sticks," many of which are legally separable from one another. Important sticks within the bundle include the powers of (a) use (including control over revenue generated), (b) exclusion, (c) security (including protection against forced transfer), (d) alienation, (e) bequest, and (f) destruction. A "full" property right would include all six sticks, with few qualifications. Licensed property rights include a weakened form of at least one stick, security against government withdrawal, and sometimes lack other sticks altogether.

Some sticks are more controversial than others, both historically and in the theoretical literature. The powers of bequest and destruction, for example, are frequently attacked in theory (e.g., Mill 1978) and weakened in practice. Testamentary disposition of land was highly restricted under feudal rules of primogeniture and entail (which specified the succession of heirs) in Europe and the United States until the nineteenth and even the twentieth century in some cases. Alienation is frequently restricted in practice as well, although less so than the related power of bequest. Of the six rights listed above, exclusion and use are the two powers most frequently considered central to ownership.[5] Although these two have also been modified and weakened at times, together they form the core of virtually any recognizable property right.

The bundle of sticks metaphor is not purely academic. In practice, fragmented powers of ownership have become quite common. For instance, legal ownership without the power of transfer in the form of life estates and other arrangements is a standard part of Western jurisprudence today (Dwyer and Menell 1998, 141). Ownership without day-to-day control has become common in the business world since the rise of the modern corporation, in which salaried employees operate the business while shareholders retain the revenue from the enterprise. The Blackstonian vision is an extreme that is rarely seen, even in the Anglo-American legal system.

Other conceptions of property have prevailed in different cultures and at different times in history. It has already been noted that common property regimes, in which many sticks are lacking from what could be called the full Blackstonian bundle, are commonplace worldwide. State or public property is another prevalent form of ownership that encompasses different powers than private property itself. (Both environmentalists and natural resource industries, for example, frequently remind us that a third of U.S. land remains in federal ownership with significantly different qualities than private property.) Moving across time rather than space, one finds that private property encompassed much stronger rights against government regulation in the United States of the late nineteenth century than it does today (Horwitz 1992). Centuries earlier, feudal property systems distributed private rights of use and transfer in very different (and more restricted) ways than the current forms of private ownership (Rose 1994, 58–61).

Added to this multiplicity of property forms are the contributions of modern scholars seeking to redefine property rights still further. Thus, the literature on property has included a host of alternative bundles. Writers have attempted to expand the notion of property to include rights both to government largesse (Charles Reich, *New Property*, 1964) and to basic human freedoms (Macpherson 1977). Another influential argument has called for a contrast between "personal" and "fungible property," describing different legal treatment under current law for each category (Radin 1982). Still another has argued that any case for private ownership must include a "property right" to basic human needs to be morally persuasive (Waldron 1993). The definition of property is contested ground, resulting in a lively debate with little agreement on what a property right may be, at least at the margins.

Given such a wide variety of definitions, the decision to describe rights created by certain market-based policies as a form of property seems quite reasonable. Licensed property created by a market-based policy is just another distinctive member of the property family. A licensed property right by definition contains many of the traditional powers of ownership but lacks others, including protection against government confiscation. Of course, in some cases licensed property rights may contain so few traditional sticks in the bundle as to defy continued description by the term *property*. The line separating such a limited licensed property right from a bare license, or effective nonownership, is not always clear. But the specific powers of ownership presented here should provide enough guidance that in most cases, what is or is not meaningfully called a property right will easily be determined. Some specific examples of licensed property in the next section will reinforce this point.

EXAMPLES OF LICENSED PROPERTY ALLOCATIONS

All effective policies protecting open-access resources must limit use in some manner.[6] Any limitations on use, however, create winners and losers:

people who get more access versus those who get less, or people who pay more for their use versus people who pay less. This is true regardless of the mechanism—public or private, command-and-control or market-based—by which access is restricted.[7] Thus, in terms of the initial allocation of use rights, issues of distribution and equity arise whether the restrictions are created via regulations with or without a market-based orientation. In light of this fact, one might ask, why should one limit oneself to considering licensed property policies? Why not focus on equity norms for *all* policy options that limit resource access and use? This section will provide four answers to that question, ranging from conceptual to methodological in nature. The answers will also explain why the acid rain, grazing, and climate change cases are especially well suited to the questions posed by this book.

The first reply is conceptual. Licensed property policies create winners and losers with greater permanence than command-and-control approaches. Rights granted to users in licensed property policies are sufficiently secure, exclusive, and well defined as to be a recognizable form of private ownership. They are designed to be more perpetual and stable than the kinds of access granted under the command-and-control model. Bureau of Land Management grazing permits, for example, are secure enough to increase the value of the permit holder's private ranch for purposes of determining mortgage values and inheritance taxes (Raymond 1997). Because licensed property rights are more permanent, they require even greater care in their initial distribution, as those legislating the acid rain program and the Taylor Grazing Act, for instance, were keenly aware. This relative permanence makes licensed property the logical starting point for a discussion of equity in environmental policy in general.

The second justification for focusing on licensed property is methodological. Studying a subset of environmental regulations that are property-based focuses the research process without rendering it trivial. Outcomes of this inquiry will be most applicable to licensed property policies but will also have relevance to other legal methods of restricting resource access and use. Licensed property policies also tend to include more explicit debate over equity issues during the allocation of access rights than their command-and-control alternatives. The relative prominence of equity issues in these cases thus facilitates study of the process.

A third justification is theoretical. Equity ideas shape public policy in a wide variety of ways. Explaining how equity norms affect public policy at any level is a daunting task; doing so on a wide range of policies is even more so. A study limited to market-based policies is therefore a useful first step in narrowing the field. Focusing on the subset of those policies creating licensed property rights is even more useful because it permits the application of a specific and flourishing area of legal philosophy as a theoretical framework. Theorists have debated entitlements to property for centuries, forming arguments and normative positions that are extremely relevant to the allocation conflicts that constitute this study. Licensed property policies allow the use of this focused body of property theory as a source of relevant norms, rather than far more diverse and less manageable ideas about distributive justice and equity writ large, in framing the debates that occurred.

Finally, an extensive array of licensed property examples also argues for the current research focus. One of the hallmarks of the current debate over market-based policies is a periodic misconception that they are new or unprecedented. In reality, governments have created licensed property rights throughout history, frequently to address environmental issues. Extensive data are therefore available for understanding how other allocations of licensed property have occurred. In addition, given the prevalence of such policies, the findings in one case are likely to have wider relevance. The examples below provide a sense of the rich diversity of settings in which licensed property rights have been created and distributed:

Unpatented mining claims. U.S. mining law provides a clear method for individuals to obtain full legal ownership of federal land containing hardrock minerals. Such private claims on the public lands date back to mining laws passed in the nineteenth century that continue to control the allocation process to this day (30 USCA sec. 22; Coggins et al. 1993). Before being "patented" and passing into private hands via a fee-simple title to the land, these claims exist in their "unpatented" form.[8] Such unpatented claims still provide an exclusive, liquid, and relatively secure form of ownership for the private holder (Graf 1997). However, they can lapse into government ownership if certain ongoing (but fairly undemanding) conditions are not met.[9] As such, they are a long-standing example of licensed property in American environmental policy.

Electromagnetic spectrum for broadcasting. The federal government initially allocated broadcasting rights in the 1920s and 1930s. The rights granted to a specific broadcast frequency are fully exclusive (i.e., use by anyone other than the right holder is prohibited) but moderately insecure. Broadcasters, for example, must perform certain public duties (such as public service announcements) and meet certain basic moral standards in their programming content to retain control over their wavelengths. These rights cannot be traded freely, although they typically accompany the sale of the broadcasting corporation possessing them. Thus, broadcast rights are a relatively weak form of licensed property: fully exclusive, transferable only in a limited manner, and subject to a greater threat of government revocation without compensation than many other examples.

Grazing permits on federal lands. Since the 1897 Forest Reserve Act, federal agencies have regulated grazing on public lands. These agencies continue to control access to public land forage through10-year permits with a preference right of renewal for current holders. Grazing permits under both the Forest Service and the Bureau of Land Management are not vested property rights, but they are exclusive rights of access to forage with a very strong *de facto* security of tenure. They are also transferable, but only as part of transactions involving the private lands to which the permits are attached. Though less secure than unpatented mining claims, they remain one of the most secure and stable forms of licensed property.

Individual transferable quotas (ITQs) for fisheries. Several fisheries worldwide have turned to licensed property management schemes for both conservation and efficiency reasons. In the United States, several programs have been created since being authorized by the Magnuson Act in 1976 (CRS 1995). Leading examples include the Atlantic surf clam and ocean quahogs program in the Northeast, and the halibut and sablefish program in Alaska. Although the specifics of each program vary, usually the ITQ entitles the holder to catch a certain percentage of the total allowable catch in the fishery each year. Most ITQ programs create a right that is fully exclusive and extremely secure—granted in perpetuity to current users, albeit only as a percentage of a floating total allowable catch figure (Macinko 1993). They are also more liquid than grazing permits or unpatented mining claims, generally being available for sale to any interested party. As such, ITQs are closer to the Blackstonian ownership ideal than most other licensed property examples.

Air pollution emissions allowances. An allowance to emit sulfur dioxide (SO_2) into the atmosphere is probably the best-known example of a licensed property right. Created by the Clean Air Act Amendments of 1990, SO_2 allowances are fully transferable to any party, including nonusers, such as environmental groups. They are also fully exclusive: each allowance entitles the holder to emit one ton of sulfur dioxide in a given year (or later), and emissions without allowances are not permitted. In one sense, the security of an SO_2 allowance is greater than the typical ITQ: holders are entitled to a specific level of use, whereas holders of ITQs are entitled only to a fixed percentage of a variable total catch figure.[10] On the other hand, SO_2 allowances remain explicitly subject to reduction or change without compensation by their statutory language, so their security still falls well short of a vested right of private property. Indeed, given their relatively short existence, as well as ongoing legislative proposals to adjust their value, one would have to call these rights less secure than federal grazing permits, which have exhibited remarkable stability over many decades. Emissions allowances for greenhouse gases contemplated in the climate change case are being modeled on the acid rain example in many respects, although at present they remain more limited in their tradability. As the product of an international treaty, however, they represent a more secure entitlement than SO_2 allowances existing at the continued sufferance of Congress.

Table 1.1 summarizes the forms of licensed property created under these five examples. Two important points emerge from the table. First, licensed property policies for managing environmental resources are not new to the late twentieth century. Important forms of licensed property have been created at various times for various public resources dating back to the nineteenth century. Second, despite the firm economic ideal for property rights, the type of right actually created varies significantly. The resulting differences in term, liquidity, and security, for example, make more apparent the flexibility of the licensed property approach as it is put into practice.

Any of the cases summarized in Table 1.1 could serve as interesting points of comparison; indeed, other examples of licensed property could be added to this group.[11] This book focuses on the air pollution and grazing examples, however, for several reasons. The clean air case is a logical point of departure, since SO_2 allowances are the most prominent and widely discussed example of a market-based environmental policy today. As a prominent attempt to bring similar licensed property rights to the international arena, the climate change negotiation process serves as an obvious point of comparison.

Perhaps more surprisingly, grazing permits present an outstanding case for a domestic comparison with SO_2 allowances. Both programs attempted

Table 1.1 Examples of Licensed Property Rights

Kind of right	*Date created*	*Resource owned*	*Length of term*	*Relative liquidity*	*Relative security*
Unpatented mining claim	1866–1872	Public land	Perpetual with conditions	Strong	Very strong
Broadcast license	1927	Electromagnetic frequencies	8 years	Moderate	Moderate
Grazing permit	1934	Public range forage	Up to 10 years	Moderate	Strong
Fishery ITQ (in U.S.)	1976	Fish stock	No fixed limit	Strong	Strong
SO_2 allowance	1990	Atmospheric sink capacity	No fixed limit	Strong	Moderate
Greenhouse gas allowance	1997	Atmospheric sink capacity	5 years	Limited (to date)	Strong (to date)

to address severe environmental problems and involved specific federal legislative actions that were considered nationally significant and politically controversial. Both also embodied contentious allocation processes for those rights that drew heavily on equity-based arguments. Previous scholarship has not connected grazing permits to emissions allowances, however, for a number of possible reasons. Although they are transferable, grazing permits are not traded to equalize the marginal costs of compliance with regulatory standards, as emissions allowances are. Advocates and students of market-based policies also tend to focus on pollution problems, especially with respect to air and water; grazing tends to be discussed by an entirely different group of scholars and policymakers. Recognition of the grazing case as an important and distinctive example of a market-based policy using the incentives of private property has therefore been lacking, a situation this study hopes to rectify.

NORMS, MARKET-BASED POLICIES, AND RATIONAL CHOICE

Scholars have increasingly recognized the importance of norms in shaping human behavior. More recently, this recognition has extended to explanations of political action. Even political scientists working on public choice theory, with its models of political behavior based on self-interest, have noted the likely influence of norms on legislative outcomes. Despite this acknowledgement, empirical work on the political influence of specific norms has been more limited. This is particularly true of research on the Clean Air Act Amendments and the Taylor Grazing Act, in which attention to the allocation process in general and the influence of norms in particular has been lacking. This section will briefly review work on public choice and market-based policies to help develop the theoretical context and justifications for this book.

Equity, Norms, and Rational Choice Theory

Rational choice theory is a model used to understand and predict human behavior. Rooted in the discipline of economics, the theory assumes in its purest form that individual behavior can be explained in terms of motivations of self-interest. Rational choice theory moved into the world of political science in the 1950s, leaping to prominence with the publication of Anthony Downs's (1957) book, *An Economic Theory of Democracy*. Since then, the growing influence of rational choice has "recast much of the intellectual landscape in the discipline of political science" (Green and Shapiro 1994, 3). Indeed, although rational choice has moved into many social science disciplines, including law (Posner 1972), sociology (Zafirovski 2000), and others, it has been argued that the movement's most significant impact beyond economics has been on political science (Miller 1997). This extension of rational choice into the political realm is still controversial, particularly because of some powerful normative arguments between certain rational choice theorists (Stigler 1971; Riker 1982) and their critics (Sagoff 1988; Kelman 1988) over the proper role of government regulation and even of democracy itself.

Models of behavior under the rational choice rubric generally agree on two assumptions: (a) individuals maximize the expected value of their actions, and (b) these expected values are calculated according to personal preferences that are internally consistent and relatively stable. Beyond this basic description, there are numerous variants reflective of the vast literature that has developed on the topic. Two contrasting types of rational choice accounts described by Green and Shapiro (1994) are of particular importance to this discussion: models using "thin" versus "thick" ideas of rationality, and those relying on internalist versus externalist explanations.

The thin model of rational choice makes no statements or assumptions about the content of an individual's preferences. It only requires that those preferences be consistently ordered—that is, if *a* is preferred over *b*, and *b* over *c*, then *a* must be preferred over *c* as well. A problem with the thin model, as Monroe (1991) has pointed out, is that it verges on tautology: if preferences are revealed by an actor's choices, it is almost impossible to prove that he or she is choosing irrationally. This quality also makes thin models very difficult to test empirically (Green and Shapiro 1994). Thick models of rational choice go a step further by adding some specific content to an actor's goals—for example, the maximization of a specific good like profits, income, or votes. This content makes the theory less tautological and more subject to empirical testing but raises the additional question of whether actors are in fact motivated by the specific preferences indicated by the theorist.

The second division in the theory is epistemological: is rational choice a description of actual human thought processes, or simply an effective predictor for the results of same? Internalist accounts of rational choice argue that people think in terms of self-interest and maximizing behavior. Externalist accounts argue that the inner workings of individual minds are both irrelevant and difficult to fathom; what matters is that the assumption of rationality accurately predicts actual behavior. Externalist accounts certainly gain support from various studies (e.g., Thaler 1991) that have shown people frequently approach economic and other choices in irrational ways. Critics have pointed out, however, that if rational choice theory is to rely on an externalist justification, then there is increased pressure to demonstrate good predictive empirical results (Green and Shapiro 1994).

Indeed, of the many critiques of the rational choice approach in politics, one of the most cogent has been that the theory is a poor predictor of actual political behavior. In particular, the theory that legislators will act to maximize their constituents' economic interests in order to remain in office has come up short empirically. Numerous studies on issues ranging from surface mining control (Kalt and Zupan 1984) to funding for Amtrak (Baron 1990) to the allocation of SO_2 allowances (Joskow and Schmalensee 1998) have found that legislative behavior is not clearly determined by any simple model based on variables of economic self-interest. The conclusion that other factors, such as ideology, play an important role in shaping legislative outcomes is echoed by a leading recent attempt to unify models of legislative behavior (Keohane et al. 1998). It is also confirmed by less quantitative studies of legislative behavior (Drew 1978) and in research on the negotiation of international agreements (Grubb and Sebenius 1992). In reality, the assumption that legislators seeking reelection will devotedly serve the economic interests of their constituents is inadequate for explaining legislative action. As Kalt and Zupan (1984, 279) conclude, "... approaches which confine themselves to a view of political actors as narrowly egocentric maximizers explain and predict legislative outcomes poorly."

This "irrational" legislative diversion from the economic interests of constituents can be explained in two ways. "Slack" in the principal-agent relationship may give the legislator room to vote according to personal ideology on certain issues without being reprimanded by his constituency. Alternatively, constituents may hold politically relevant ideological or altruistic ideas as well as economic interests, which are then reflected faithfully by the legislator's voting behavior. Research seems to indicate that the second explanation better describes actual legislative behavior than the first (Kau and Rubin 1993; Peltzman 1984). Either way, legislative behavior clearly appears to be determined in part by noneconomic factors, such as ideology. In addition, the ideologies in question tend to center on ideas of equity (Kalt and Zupan 1984; Grubb and Sebenius 1992). Thus, the empirical literature regarding legislative behavior retains a place for ideological principles of equity, despite the attempt by some theorists to rely exclusively on the self-interest of politicians serving the economic demands of their constituencies.

These empirical findings notwithstanding, many supporters of rational choice remain undeterred, arguing that "it takes a theory to beat a theory." Although rational choice may not be perfect, they maintain that it performs significantly better than any other theory to date in explaining human behavior (Elster 1986). Yet some researchers are seeking to add information to rational choice models in order to improve their empirical performance (Almond 1991). In a similar manner, an important goal of this book is to make a contribution toward developing a much "thicker" version of the rational choice approach.

In particular, integrating social norms into the rational choice model has drawn significant attention from some scholars. As noted in Chapter 1, a norm is generally defined as a socially created rule of behavior that prescribes or proscribes certain actions, often counter to the immediate self-interest of the actor following the rule. Norms are not rules of law; they are social rules that control human behavior outside the legal apparatus of government. Ellickson (1991) has written persuasively on the role of norms in regulating individual behavior regardless of the legal rules involved, and Elster (1986, 1990) has noted that norms provide a powerful alternative to rational choice theory for explaining some human behaviors. Noll and Weingast (1991) have gone further and argued that norms may shape the goals pursued by rational actors, in effect providing some of the specific content of personal preferences in thick versions of rational choice theory. Their work emphasizes the role of norms based on ideas of fairness. In this respect, social norms seem to be a powerful way of describing individual goals in a manner that moves beyond simple self-interest into something more intuitive and empirically persuasive.

A fairly extensive literature documents the existence of equity norms in politics and society. A few scholars have tried to demonstrate the influence of ethical values on certain policy decisions (Craig et al. 1993; Bazerman 1985; Zajac 1978). In addition, a more extensive body of experimental work in political science and game theory has produced numerous outcomes in which subjects deviate from self-interested behavior in favor of egalitarian or other equity-based norms (Eavey 1991; Surazska 1986; Hoffman and Spitzer 1985; Eavey and Miller 1984; Miller and Oppenheimer 1982; Nydegger and Owen 1977). Such results remain hard to generalize, and some studies contest the apparent influence of equity norms presented in these works (Rutstrom and Williams 2000; Weingast 1979). In addition, many of these studies find that support for "fairness" norms varies in different situations, making generalization even more challenging. Nevertheless, significant deviation from purely self-interested outcomes in favor of those based on ideas of fairness and equity are well documented in these branches of political science and economic research.

Much work remains, however, on creating better theories of norms to avoid simple *post hoc* explanations of specific behaviors. Noll and Weingast (1991) avoid this difficulty by restricting the possible content of such norms to certain commonly promulgated philosophical positions, as is done in this study. Another extensive literature, largely in the disciplines of psychology and sociology, seeks to document which established norms of equity are common among individuals in society (Miller 1992; Messick and Cook 1983; Cook and Hegtvedt 1983; Hochschild 1981; Walster and Walster 1975). Yet Miller

(1992) observes that there is only a limited interaction between these empirical studies of equity norms and the traditional philosophical works referred to by Noll and Weingast (1991). This book seeks to increase those connections by scrutinizing traditional property theory for the content of norms at work in the creation and allocation of licensed property rights.

Equity, Norms, and Market-Based Policies

The literature on market-based environmental policies dates back to the early twentieth century, with Pigou (1932) being commonly cited as an initial source. In 1960, economist Ronald Coase revolutionized the field with his argument in favor of clearly defined property rights as the key to reducing the social cost of regulation.[12] Subsequent thinkers drew on the literature to push for market-based regulations, including emissions taxes as well as tradable permit schemes (Dales 1968; Montgomery 1972; Kneese and Schultze 1975; Schultze 1977), and others emphasized the ability of well-specified private property rights to efficiently prevent environmental damage (Posner 1972; Tietenberg 1988). As the U.S. Environmental Protection Agency gradually began experimenting with market-based approaches following the Clean Air Act Amendments of 1977, commentators concentrated on the air pollution case (Hahn 1988, 1989; Dudek and Palmisano 1988; Ackerman and Stewart 1988; Liroff 1986; Tietenberg 1985). After celebrating the passage of the acid rain program in 1990, scholars began advocating an expansion of market-based policies to other problems, including waste disposal (Hockenstein et al. 1997), greenhouse gas emissions (ELI 1997), and even allocation of geostationary orbits for satellites (Macauley 1997).

In the vast majority of those works, regulatory efficiency serves as a dominant policy goal.[13] This goal manifests itself in three themes throughout the literature. The first is the need for strong property rights. Market-based policies work best, according to this argument, when the property rights in question are as exclusive, secure, and transferable as possible. Weakness in one or more of these powers of ownership is frequently criticized as the reason for an undesirable policy outcome. The second theme is the framing of a regulatory problem as a "market failure," or an issue of externalities, for which market-based policies are required to ensure that the full effects of a resource user's actions are included in the action's price. A classic externality example is a factory that pollutes the air and dirties a neighbor's laundry hung out to dry. Without a clear assignment of property rights to clean air (á la Coase), the factory owner does not include the negative impact of dirtying the laundry in his decisionmaking, leading him to emit an inefficiently large amount of pollution.[14] The third theme is the assumption of the self-interested actor. Whether explicitly argued for or implicitly included as a part of the model, the assumption that individuals subject to regulation can be presumed or even encouraged to act in a selfish manner is central to these works.

Those three themes in the literature have tended until recently to make equity an afterthought. A focus on efficiency as a measure of policy success leaves little room for the consideration of equity issues. The presumption in favor of strong property rights makes the most sense from this efficiency perspective (despite the regular weakening of such rights in practice, as summarized in Table 1.1). And the view of environmental problems as externalities also omits equity concerns. Put the right marginal cost on the polluting action (i.e., "get the prices right"), and one has solved the problem according to this account, regardless of the distributive implications of the policy. Finally, the assumption of the self-interested actor also pushes equity arguments aside. If all actors are assumed to be exclusively self-interested with no concern for others, then the public policy goal of making a fair or equitable policy begins to seem ill-conceived or even naïve.

Of course, some have argued that equity *should* matter more in discussions of market-based policies. Steven Kelman (1981), for example, is frequently cited for providing an explicitly normative argument against certain market-based instruments in environmental policy. According to Kelman, the use of economic incentives may lessen the stigma attached to polluting behavior and thereby create a society with values that are socially undesirable. The popular press, some environmental advocates, and even occasionally the academic literature have featured arguments critical of the "right to pollute"—arguments that owe a great deal to Kelman's work.[15]

Other critics have focused more generally on the distributive impacts of the efficiency-driven perspective. A few, including Nobel prize–winning economist Amartya Sen (1992), have deemphasized efficiency within economics and public policy by locating it within a broader set of social concerns, distributive justice among them. Others have written eloquently of the need to consider equity rather than simply efficiency in environmental policy settings (Shue 1993; Rees 1990; Sagoff 1988). Still others have urged a greater consideration for equity with respect to market-based initiatives, particularly those contemplated on an international level (Rose 1998; Hays 1995; Albin 1995; Rose and Stevens 1993). Lately, some of these arguments appear to have hit home, as there is a welcome and growing trend of works paying greater theoretical attention to equity implications (e.g., Sterner 2002; Portney and Weyant 1999).

Indeed, the climate change case is a leading example of this growing trend, in part because of the international aspects of the problem. The importance of shared principles of fairness in international negotiations is widely recognized, even in the environmental context (Albin 2000; Rose 1998; Parson and Zeckhauser 1995; Grubb and Sebenius 1992). As a result, a large number of authors have offered various principles of equity to serve as a basis of any future climate change agreement (Ott and Sachs 2000; Shue 1999; Burtraw and Toman 1992; Rose 1990; Chapman and Drennen 1990). Many have suggested allocations that converge in the long term on an equal per capita distribution of emissions rights globally (Sagar 2000; Baer et al. 2000; Meyer 1999; Agarwal and Narain 1991). Others have proposed alternative allocation rules, including those based on a nation's energy efficiency (Solomon and Ahuja 1991), ability to pay (Pew Center 1998a), or historical responsibility for the problem (Smith 1996).

These works make a strong case for the need to consider equity in public policy, especially in the international context. They tend to protest market-based instruments and the unbridled pursuit of efficiency on philosophical grounds, however, rather than considering how equity ideals already influence policy in practice. In fact, there is little empirical work on how equity concerns already affect market-based policies or might be expected to do so in the future. Indeed, there is little empirical work of any sort on the process of distributing licensed property rights under market-based policies, whether for acid rain, public lands grazing, or climate change.

Most discussions of existing licensed property policies pause at the initial allocation just long enough to grimace at the untidy nature of the process. Those that address the issue of distribution often portray the allocation process as something of a headache for policymakers (Solomon 1995; McLean 1997), one frequently solved by grandfathering the rights to existing users based on the status quo (Tietenberg 2002; Stavins 1997; Burtraw and Swift 1996; ELI 1997). Alternatively, scholars use the allocation process as an example of rent seeking and "capture" of the regulatory process by special interests (Joskow and Schmalensee 1998; McConnell 1966; Foss 1960). Of the writers who specifically recommend an alternative method of allocation for a licensed property policy, most support an initial auction of rights by the government, with funds raised being used to offset inefficient forms of taxation (Burtraw 1998; Hahn and Noll 1983).

This pattern certainly holds true for most research on the acid rain program. Scholars frequently note the program's success in reducing SO_2 emissions[16] and have considered in detail the program's various efficiency implications (Dudek et al. 1997; Rose 1997). However, although countless studies have analyzed the efficiency and ecological effectiveness of the program (Ellerman et al. 2000; Burtraw 1998; Stavins 1997; Bohi and Burtraw 1997; Burtraw and Swift 1996; CRS 1994; GAO 1994; Bernstein et al. 1994), almost none have considered its relationship to equity (but see Hays 1995). Only two studies to date have considered the actual process of allocating allowances under the Clean Air Act Amendments in significant detail (Joskow and Schmalensee 1998; Kete 1992a), and neither prioritizes the role of equity or the influence of norms.

The omission is remarkable, given that the policy created and distributed an extensive new kind of wealth in the form of SO_2 allowances. Joskow and Schmalensee (1998, 39) even go so far as to call this lack of attention to the distributive aspects of market-based policies a "serious gap in the literature." Allowances are assets with significant value; those used by U.S. utilities in one year alone are worth billions of dollars at current market prices. The equity questions raised by such a distribution of wealth are profound, and they were a major part of the policy debate regarding the passage of the bill. Yet scholarly work has been sparse on this aspect of the acid rain program.

The treatment of the distribution issue in the public lands grazing literature is similarly limited. Dominated by questions of administrative behavior, agency capture by private ranchers (Klyza 1996; Culhane 1981; Voigt 1976; Foss 1960; Calef 1960), and below-market grazing fees (Coggins et al. 1993, 702–704), the grazing literature lacks the extensive references to market-based policy found in the air pollution case.[17] There are some works supporting the creation of vested private property rights on the public range in pursuit of economic efficiency (Carney 1998; Nelson 1986; Gardner 1984). Other writers have followed Harold Demsetz (1967) by documenting the historical emergence of extralegal property rights on public lands as scarcity increased and technology improved (Dennen 1976; Anderson and

Hill 1975). Still others, including those sympathetic to the Wise Use movement, have argued that federal grazing permits already represent fully vested legal rights of ownership, based in part on these nineteenth-century practices and customs (Falen and Budd-Falen 1993; Hage 1989).

Although concerned with property rights, none of the entrants in the grazing literature seek to understand how equity ideas actually shaped the allocation of licensed property in 1934–1938, a process that remains important to understanding public range controversies to this day. In fact, there is no detailed accounting of the early years of range administration under the Taylor Grazing Act. Merrill (2002) provides a narrower history of federal grazing policy, but lingers only briefly on the TGA allocation process. Peffer (1951) is the standard authority on this topic, but her book covers such a large swath of time and resources (range, timber, minerals) that she spends less than 20 pages on public domain grazing administration in the 1930s. Perhaps more remarkably, the canonical public lands history by Gates (1968) devotes only 5 of its 600 pages to the same period. The two studies that do pay significant attention to the allocation of federal grazing permits cover the Forest Service rather than the Department of the Interior (Rowley 1985; West 1982). Detailed historical coverage of the early allocations under the Taylor Grazing Act is therefore limited to former Division of Grazing Director Ferry Carpenter's (1984) personal recollections (which constitute only one chapter of his book) and another Grazing Service employee's personal memoirs (Klemme 1984). This relative lack of scholarly attention is unfortunate because the 1934–1938 period was a fascinating time of an intense struggle over resource allocation on the public domain. The tale is well worth telling in greater detail.

Perhaps more surprisingly, despite the greater attention to equity issues in the climate change case, there is still little work that seeks to understand *how* equity ideas have in practice played out in the process so far. Although several outstanding volumes describe the negotiation of the Kyoto Protocol, none of them devote extended time to the specific role played by ideas of equity or property norms (Oberthur and Ott 2000; Grubb et al. 1999). A partial exception is David Victor's 2001 book, *The Collapse of the Kyoto Protocol*, which explicitly considers the challenges of the allocation process under the Kyoto model. Victor's agenda is different from the one pursued here, however. His discussion focuses on the difficulty of any allocation of emissions allowances potentially worth trillions of dollars through international negotiations in which any nation can effectively veto the outcome (Victor 2001). Although I agree that emissions allocation is the Gordian knot of climate change policy, I reach that conclusion by a different path, focusing on the potentially irreconcilable property-based arguments being made by various nations. In short, although prescriptions for equitable greenhouse gas emissions allocations are prevalent, empirical analysis of the specific normative principles shaping the allocation process are largely absent.

Some of the neglect of the allocation issue, especially in terms of empirical research, may be attributable to the ongoing influence of Coase. By demonstrating that *any* initial allocation of rights results in an efficient final distribution (assuming perfect information and zero transaction costs), Coase rendered the issue irrelevant from a perspective initially endorsed by many economists. His continuing influence is well demonstrated by a 1997 study of emissions trading by Congress, which concluded that "the actual method for dividing up the total allocation of emissions is irrelevant to making a tradable emissions market successful" (Joint Economic Committee 1997). Although later writers have noted that high transaction costs might make the initial distribution more important by limiting trading, their concerns remain steeped in a desire for an efficient outcome. Ironically, as Tietenberg (1992) notes, the Coase Theorem could be used to defend an initial distribution shaped primarily by *equity* principles, since it concludes that all initial distributions of rights will result in an equally efficient outcome! Unfortunately, since a majority of supporters of property-based policies still seek efficiency and ecological improvement as their primary policy goals, the theorem has served instead to alleviate the need for consideration of the allocation process at all.

This book investigates the initial allocation of rights in a manner that goes beyond the Coase Theorem. To consider the efficiency of a licensed property policy, the initial allocation may well be marginal. But to understand the role of fairness in shaping that policy, there is a growing recognition that the initial distribution of rights is central. Despite the inability of traditional rational choice models to fully explain political behavior in cases like these, the ability of norms based on property theory to explain these allocation struggles has never been considered as it will be here. The results will support the growing role of norms in theories of rational choice while filling a gap in the empirical literature regarding market-based policies in general and the specific cases under investigation.

OBJECTIONS TO STUDYING EQUITY IN PUBLIC POLICY

Despite the growing interest in norms, there remains resistance in some quarters to the study of equity in public policy. Many scholars and professional analysts of public policy issues still tend to shy away from distributional and other policy questions that raise difficult normative issues (Young 1994; Amy 1984). Typically, this aversion is justified by either a normative or a pragmatic objection. The normative argument defends an efficient policy outcome as the most ethically justified goal for public policymakers, thereby rendering consideration of other equity-based outcomes largely irrelevant. The pragmatic argument, by contrast, notes the importance of equity concepts (other than efficiency) in policymaking but declines to consider them further because they are too difficult, murky, or inconclusive to discuss and research productively. Sometimes these arguments are made explicitly, and sometimes they are implicit within policy discussions. Either way, since this study is dedicated to understanding the role of equity ideas in specific policy situations, a response to these overarching objections is warranted before moving into the body of the argument.

The Normative Objection: Efficiency Is the Preferred Goal

Research on public policy today relies heavily on economic analysis. Economics in turn focuses on the concept of efficiency, for several persuasive reasons. Different measures of efficiency are relatively easy to quantify, making powerful mathematical analysis and modeling of different policy problems possible. In the simple sense of "getting the most bang for the buck," efficiency is a policy goal that is hard to oppose, all other things being equal. The Arrow Theorem (Arrow 1951), on the impossibility of a unique democratically chosen social welfare function by which one could evaluate various distributive outcomes, has also had an important influence in this regard. Finally, the search for the most efficient public policies (rather than the fairest, or the most desirable by some other measure) enhances the "nonpartisan technical expert" image sought by many professional policy analysts and bureaucrats (Amy 1984; Tribe 1972). Thus, the pursuit of efficiency and the use of quantitative economic methods are now a standard part of the public policy system, with leading public policy textbooks, for example, supporting this approach (e.g., Munger 2000; Stokey and Zeckhauser 1978).

Of course, different definitions of *efficiency* imply different policy objectives. The least controversial definition, at least on an ethical basis, is probably Pareto-efficiency. Named for Italian economist Vilfredo Pareto, a Pareto-efficient situation is one in which it is impossible to make anyone better off without making at least one other person worse off. Similarly, a Pareto improvement is a change in resource allocation that makes at least one person better off while not reducing the well-being of anyone else. In this sense, a policy seeking efficiency via Pareto improvement might be more informally described as pursuing a win–win solution, making it difficult for anyone to oppose. However, Pareto improvements are relatively rare in politics—most public policies have winners *and* losers. Seeking only Pareto-optimal results would severely limit the choices of policymakers.

An alternative, and more influential, definition of *efficiency* in public policy scholarship is the Kaldor-Hicks version. Named after English economists Nicholas Kaldor and John Hicks, Kaldor-Hicks efficiency is the foundation of cost-benefit analysis and most other techniques that dominate public policy analysis today. The idea expands efficiency to include outcomes that *could* be Pareto improvements if the winners compensated the losers. It is crucial to note here that such compensation need not actually happen; to meet the Kaldor-Hicks standard, it is only required that it theoretically *could* happen in a manner leading to a Pareto outcome. Obviously, this is a far less restrictive notion of efficiency that permits many more public policy actions.

How might one support Kaldor-Hicks (K-H) efficiency as a preferred policy goal on normative grounds? One of the best such arguments is presented by Herman Leonard and Richard Zeckhauser. In a thoughtful and wide-ranging defense of cost-benefit analysis as a public policy tool, Leonard and Zeckhauser (1986) argue that impartial individuals would give their hypothetical consent to policies with K-H–efficient outcomes as long as they did not know their status as winners or losers in advance. Requiring actual Pareto improvement would be far too restrictive, they argue, since providing compensation to losers may be impractical (or even undesirable, in terms of encouraging them to overstate their damages in hopes of greater compensation). Although society might actually compensate losers for certain extraordinary individual harms, they conclude that in general we are better off letting most small wins and losses due to public policy cancel out, more or less, over the long run. Summing up their argument, they claim to know of "no other mechanism for making [public policy] choice that has an ethical underpinning" (1986, 33–37).

However careful and well considered, this defense of K-H efficiency may rely a little too strongly on a single "silver bullet"—hypothetical consent. Although it is plausible that individuals would agree to policies seeking K-H–efficient outcomes, it seems at least equally plausible that they would not agree. Indeed, John Rawls (1972) famously relies on a similar notion of hypothetical consent in *A Theory of Justice* to support an entirely different standard that maximizes the distributive share of those who end up the worst off. Other scholars, in turn, have criticized the belief that individuals would necessarily agree to Rawls's standard (Waldron 1986). Determining just what kinds of distributive rules people would support hypothetically is a tricky task, it appears, and depends heavily on what assumptions one makes about their degree of risk aversion and other factors. Thus, hypothetical consent serves as a relatively weak normative basis for seeking K-H efficiency over all other outcomes in public policy.

More generally, the pursuit of K-H efficiency can be supported by a utilitarian ethic. Utilitarianism, of course, is the moral philosophy that, at its most basic level, insists on achieving the "greatest good for the greatest number" (Mill 1998). A mandate to seek K-H–efficient outcomes also focuses on maximizing the size of the "pie" for society and strongly resembles this utilitarian maxim. Indeed, more than one observer of K-H efficiency's powerful influence has noted its direct connection to a utilitarian outlook on the world (Kelman 1992; Amy 1984). Strictly speaking, K-H–efficient outcomes may not generate the greatest "good" in society—efficiency being typically measured in tangible units of wealth like dollars, whereas utility (or happiness) is a more intangible concept resisting cardinal measurements. A policy creating an additional dollar of wealth in society, in other words, may well not create an additional unit of utility. But in general, many public policy scholars seem willing to ground their pursuit of efficiency as a controlling public policy goal at least informally on a utilitarian outlook (e.g., Munger 2000; Stokey and Zeckhauser 1978).

The difficulty with their argument is similar to the problem with the hypothetical consent gambit: though a plausible candidate, utilitarianism hardly dominates the intellectual space of ethics and political philosophy to the exclusion of other perspectives. Indeed, as Steven Kelman (1992) has pointed out, at this point utilitarianism is in a minority position in moral philosophy, attracting more critics than supporters. Prominent moral philosophers have made powerful critiques of the utilitarian outlook, noting its demanding requirements of beneficence and personal sacrifice, as well as its apparently counterintuitive dictates in many moral situations (Williams 1998; Kelman 1992). Utilitarianism remains an important part of some important modern theories of ethical public policymaking (e.g., Goodin 1982), but any normative claim that it merits an *exclusive* influence over policymakers seems to overreach, in light of the many serious nonconsequentialist and pluralist alternatives that exist (Gillroy 2000; Anderson 1993; Sagoff 1988; Anderson 1979; Hampshire 1978).

Thus, both the hypothetical consent and the utilitarian arguments fail to exclude other visions of equity from the study and analysis of public policy. There are, of course, good normative reasons for seeking K-H–efficient outcomes in public policy, but in many cases there are even more persuasive normative reasons for seeking other, less efficient alternatives. In other words, the move from efficiency as an *important* standard of public decisionmaking to the view that it is the *ethically dominant* standard seems to go too far. Given the continued controversy over K-H efficiency and utilitarianism in general, it seems quite implausible to neglect the study of other equity-based concepts in public policy on normative grounds.

The Practical Objection: Equity Defies Rigorous Analysis

"Probably the most common argument used to exclude ethics from policy analysis," writes Douglas Amy (1984, 575) in a leading journal of public policy studies, "is the contention that moral and value judgments are completely subjective and therefore not amenable to rational analysis." Amy goes on to note variations on this complaint: that moral concepts are too abstract to be relevant to public policy decisions, or that a more sophisticated normative approach would simply require too much time and energy. The effort of many public policy experts and scholars to be perceived as objective again plays a role here, since dabbling in the realm of equity norms is often seen as putting the analyst on the wrong side of the epistemological fence between facts and values (Amy 1984). Although equity ideas might matter in theory, this position concludes, they are far too difficult and impractical to include in scholarly treatments of public policy.

An excellent and more recent example of the pragmatic objection is an article by leading environmental economists Don Fullerton and Robert Stavins (1998) published in *Nature,* entitled "How Economists See the Environment." In a brief essay, the authors attempt to rebuff four "myths" about their discipline and its view of public policy. One of the myths is that economists care only about efficiency and not distribution. The authors' response eschews the normative justification above for a more pragmatic reply: economists tend to focus on efficiency because efficiency issues are more precise and unambiguous. "What constitutes an improvement in distributional equity," Fullerton and Stavins (1998, 434) continue, "... is inevitably the subject of considerable debate." Their implication is that questions of equitable distribution are best considered elsewhere. Wherever "elsewhere" may be, however, it is clearly not within the world of environmental policy, which is not the "best place" to achieve distributive goals in society (Fullerton and Stavins 1998, 434). This aversion to considering distribution in specific policy settings is hardly unique; Leonard and Zeckhauser (1986) make a similar argument, as do others.

Although the pragmatic objection is well taken, there is more to be said here. First, market-based policies frequently do have important distributive results, whether one considers them explicitly or not. Studying environmental policy without directly considering those issues does not make them go away, as public policy decisionmakers are keenly aware. More importantly, equity issues play a critical role in the actual practice of policymaking. Politicians, rather than academics, create and implement environmental policies, and politicians are quite concerned with the distributional implications of their actions. So regardless of one's beliefs about the practicality or desirability of addressing larger distributive issues in a specific policy setting, the data indicate their importance in practice. Given this situation, it seems worthwhile to understand (and even influence) the political process of distribution more deeply by integrating normative concepts into public policy analysis and scholarship.

In fact, I suspect that many of those making the practical objection would agree with this line of thinking, at least to some degree. Despite their inclination to downplay certain distributive issues, for instance, the authors of the *Nature* piece also acknowledge the need to combine efficiency and distributive issues in "a unified analysis" (Fullerton and Stavins 1998, 434). Like many other writers, however, they decline to offer specific ideas about how to craft such a synthesis. To criticize them specifically for this omission is unfair; any such description lies well outside the stated agenda of their brief essay. But the omission is symbolic, perhaps, of a larger gap in public policy scholarship. The reality is that equity issues remain the lonely wall-flowers of the literature on public policy—frequently noted languishing on the periphery of the action but rarely invited to the intellectual dance. Many analysts dutifully note the equity implications of a policy option like licensed property, but far fewer offer much guidance regarding those issues. Although this trend has recently experienced a modest reversal, particularly with regard to climate change, attention to the *empirical* role of equity in shaping actual policy outcomes remains limited. Equity issues, one is left to conclude, must be too difficult, too abstract, or simply too messy for we scholars to consider in greater depth. But difficult as they are, equity issues are central to the actual public debate over market-based proposals. Surely it is a poor reason to neglect such issues simply because talking about them in a rigorous or academic manner is difficult.

In summary, arguments against considering equity are ultimately unconvincing. There is no persuasive evidence that equity arguments should not, need not, or cannot be studied as part of a public policy research agenda. Indeed, this chapter has provided the ideas and the background necessary for presenting just such a study, focusing on the distribution of licensed property rights under market-based environmental policies. Having considered these final objections, we can at last explore how equity has affected market-based environmental policies in practice, and in what manner and under what conditions it might be expected to do so in the future. Hard as the task may be, politicians and citizens regularly debate, revise, and reject policies in terms of equity. To be policy relevant, the academic community must pay more attention to this part of the process, as scholars have already begun to do. The potential difficulty of the task is no excuse.

CHAPTER 2

A PROPERTY THEORY FRAMEWORK

Thus the Grass my Horse has bit; the Turfs my Servant has cut; and the Ore I have digg'd in any place where I have a right to them in common with others, become my Property, without the assignation or consent of any body.

—*John Locke, Second Treatise of Government*

Despite being based on norms of fairness, political arguments about the distribution of licensed property rarely refer to philosophical schools of thought or cite famous theorists by name.[1] Yet such equity claims draw consistently, if implicitly, on justifications for property that have been defined and defended by theorists for centuries. The purpose of this chapter, therefore, is to consider in more detail the property theories that help shape an initial allocation of licensed property.

Despite its theoretical focus, the goal of this chapter is not to justify any particular vision of property on normative grounds. Instead, the chapter aims to describe faithfully (and with a somewhat critical eye) several prominent understandings of ownership and the allocation rules they imply. This task is important because the equity arguments in the Taylor Grazing Act, Clean Air Act Amendments, and Kyoto Protocol allocations so strongly reflect the property ideas described here. To understand these allocation processes more clearly, it will be helpful to make the underlying theories and the conflicts between them more explicit. Thus, these ideas are useful here not because they represent how society should view property but because they are accurate and powerful descriptions of the rules and rhetoric that constitute this study's empirical results.

There will be four steps to the analysis. First, the chapter will build a general framework of property theories, focusing on the type of right created and the justification for ownership offered. A clear description of four conflicting views of property—the intrinsic and the instrumental theories central to Chapters 3 and 4 as well as the possessory and egalitarian views so vital to Chapter 5—will emerge from this discussion. Second, the chapter will review the specific allocation rules implied by each theory. Because the allocation rules are so closely related to a particular understanding of property, they can be located in the same framework as the theories themselves. Third, the chapter will present an alternative vision of ownership based on the ideas of G. W. Hegel. The Hegelian theory of property seeks to unify the conflicting ideas of the intrinsic and instrumental perspectives. As an attempt to reconcile these opposing views, the Hegelian perspective is crucial to understanding the grazing and acid rain cases and may have important implications for the creation of licensed property rights in other settings as well. Finally, the chapter will present a brief discussion of Kantian property theory, which represents a promising but still incomplete reconciliation of the possessory and egalitarian views of particular relevance to international climate change negotiations.

FOUR PERSPECTIVES ON PROPERTY THEORY

Any discussion of property theory must start by clarifying just what *property* means.[2] This work will follow the view that property, technically speaking, is not the object controlled by the owner. Rather, property is best defined as a social relationship giving an owner power over other individuals that restricts their control or use of an item or resource (Macpherson 1978). This view of property as "a right, not a thing" has become especially important as the notion of ownership has expanded over the past century to include more intangible forms. Any definition of property that was restricted to things would be strained to accommodate modern ownership rights over ideas, revenue streams, or specific regulatory expectations.[3] In addition, the view of property as a right brings its social implications front and center—the power of ownership may cause significant injury to others.

When property is viewed in this way, two significant questions immediately arise: what kind of right is a particular claim of ownership, and what conditions justify assigning that claim to a particular individual? The first question seeks to delineate the limits of an owner's power by describing the *type* of property right supported by the theory. The second question aims to clarify the *justifications* that entitle a person to the powers of ownership, whatever those powers entail. Any complete theory of property must address these two issues.

More importantly, the answers to these two questions can help sort property theories into several broad categories. The type-of-right question recalls the bundle of sticks metaphor for property from Chapter 1: ownership can entail any number of specific powers, or "sticks," including exclusion, tradability, use, and destruction. In the context of allocating licensed property, however, an especially crucial factor is the right's degree of security against partial or total confiscation by others. The discussion here will therefore focus on how *secure* or *insecure* a right each theory creates. The type of right can also vary according to whether the right is seen as *political* (a creation of government) or *prepolitical* (a "natural" right). The justification for ownership includes two crucial variables as well. Some theories justify property entitlements in terms of protecting *individual* interests, others in terms of serving general *collective* goals. In addition, some theories justify *significant redistribution* of ownership rights from the status quo; others say property should *not be redistributed*. This section will review the type and justification questions as a foundation for classifying the four specific property theories that follow: possessory, intrinsic, instrumental, and egalitarian.

The Possessory View: David Hume

Private property in this account results from a general agreement to increase individual wealth for all. In service of this goal, property rights are created to reduce barriers to trade and the cost of protecting one's holdings. Seeing property as a protection against theft and an aid to voluntary exchange, this theory emphasizes security through strong rights. Because it bases property rights on a voluntary consensus among all members of society, the theory insists on an initial allocation that is a Pareto improvement on the status quo, making at least one person better off without worsening the conditions of anyone.

The ideas of the eighteenth-century Scottish philosopher David Hume (1978) are representative of this view. Hume argues that before the creation of property, the insecurity of an individual's possessions is the biggest obstacle to increasing social wealth. In Hume's view, individuals have a perpetual selfishness that initially obstructs social cooperation. Over time, however, people realize that mutual restraint from seizing one another's possessions might work to the benefit of all, regardless of the amount of wealth each currently possesses. In essence, individuals within the Humean system see the possibility of what Jeremy Waldron (1994, 86) calls a "peace dividend" that can be shared by everyone if property rights are widely respected. This Paretian dividend is maximized by the formation of clearly accepted rules defining and protecting private ownership (Hume 1978).

Generated by collective agreement for mutual benefit, the Humean right to property is clearly a political one. "Our property," concludes Hume (1978, 491), "is nothing but those goods, whose constant possession is establish'd by the laws of society ..." This theory does not seek to recognize a "natural," or prepolitical, qualification for ownership, as for example does Locke's account. Nevertheless, it does seek a strong ownership right that maximizes the security of the property holder, protecting him or her from future involuntary losses. The Humean view of property also provides the owner a full complement of powers, including those of "prescription" (permitting new claims of ownership by use when the original ownership claim becomes unclear), "accession" (permitting the owner of an item to also own its products, such as the fruit from a tree), and unrestricted transfer (Hume 1978, 507–509, 514).

Hume's defense of strong property rights seeks to benefit society at large rather than specific individuals. Yet the Humean account rejects coercive redistribution of wealth. Instead, it grants ownership rights based on possession—whatever goods people control at the time the property system is created. Hume justifies this principle both practically (in terms of the mutual agreement required for property to arise) and on utilitarian grounds rooted in what is now called the endowment effect (e.g., Sunstein 1993). We are more attached to those items we already possess, Hume contends (1978, 503), than to those we lack, and therefore this decision to allocate according to current possession will seem "the most natural" choice. This is a pragmatic, wealth-maximizing argument for property that stresses the security of individual holdings in the interests of society at large. In this respect, the secure possession view of Hume anticipates several ideas advocated by modern neoclassical economists in the tradition of Coase (1960).[4]

The Intrinsic View: John Locke

The defender of the intrinsic view sees property as an equitable protection for the individual against society. It is a right, or "trump" in the sense made popular by Ronald Dworkin (1977), that must be respected by others in defense of the individual's personal autonomy. It is also a *prepolitical* right that exists independently of the government and must be respected by any legitimate political process. Unlike Humean theory, then, the intrinsic perspective includes the idea that property is a natural right, derived from a higher source than government that no just law can undo.

Defenders of the intrinsic position may support different standards for justifying a particular private right of ownership, although many favor some sort of unilateral action, such as personal labor (Locke 1960; Nozick 1974). Intrinsic theorists also typically advocate a secure right to do with one's property as one wishes without the threat of government interference. However, beyond a basic level of security, neither the specific sticks in the bundle nor the specific rule for assigning them is the core issue. What matters most is the basic prepolitical justification for secure ownership. Whatever the mechanism for creating valid ownership rights, and whatever those rights specifically entail, private property must be respected by government to the benefit of the individual owner at nearly all costs.

John Locke provides the most famous example in Western thought of an intrinsic property perspective. In his seventeenth-century political writing, property serves as a safeguard protecting the free-holding citizen against the potentially despotic state (Locke 1960, sec. 87). For Locke, property is a natural right that individuals derive from ownership of their bodies and their labor (sec. 27). (That people own their bodies and their labor is taken as self-evident.[5]) Everything not owned in nature is free to be appropriated by human actors, as the passage from Locke at the beginning of this chapter makes clear. By mixing the sweat of labor, either directly or through the "owned" labor of servants or slaves, with the object desired, a person becomes the owner of that item. No government action is required, except to respect the ensuing ownership claim.

Initially, Locke's system of individual appropriation is restricted in two ways. The first is the Lockean proviso that "enough and as good" must remain for others after an appropriation has taken place for it to be valid. Thus, taking ownership of common resources under conditions of scarcity is not justified simply by a unilateral action of applying one's labor; the consent of other actors is then required (Locke 1960, sec. 35).[6] The second restriction limits ownership to subsistence needs only. The natural process of spoilage and waste prevents accumulation of property beyond what one can immediately use (sec. 36). With these two restrictions, however, the Lockean account loses much of its power to justify individual appropriation. To avoid these limits, Locke modifies his argument. He evades the first restriction by emphasizing the surplus (in his lifetime) of unused land in areas like America, where he believed an abundance of resources remained available for the taking. He dispenses with the second by noting the creation of money. Money serves, according to Locke, by "implicit agreement" among men as a permanent repository of value, thus permitting unlimited accumulation of wealth without threat of spoilage (sec. 48–51).

Both the Lockean proviso and the implicit agreement to permit unlimited accumulation have been the subject of extensive subsequent commentary and critique (Munzer 1990; Waldron 1988; Becker 1977; Nozick 1974; Macpherson 1962). Nevertheless, despite its more controversial aspects, the basic Lockean system of property as a license for unlimited individual accumulation via unilateral action has held a powerful place in the American pantheon of political thought since the Revolution (Hartz 1955). Even today, the notion of ownership based on "moral desert," particularly through personal labor, remains a powerful influence on property law and social custom. As an explanation and defense of private property, this model remains central more than 300 years after the initial publication of Locke's work.

The Instrumental View: Morris Cohen

To an instrumentalist,[7] property is simply a human institution created to further the equitable ends of society. It is therefore subject to change to meet evolving social goals. The view flatly rejects that property is somehow a natural or prepolitical right. Property rights may serve in part as a bulwark against tyranny, as in the intrinsic account, but they are far from inviolate.

Property is instead a construct of government and exists at the continued pleasure of the political system. The instrumentalist supports changing public priorities by adjusting the powers of ownership and even redistributing privately owned resources over time. Another way of characterizing the instrumentalist view is to think of a property relationship creating a duty in others, rather than a right for the property holder (Hohfeld 1919).[8] As a burden on others, property is better seen as a privilege than as a right—something that must meet the ends of society as a whole to be acceptable.

Although the instrumentalist perspective is broad enough to encompass a wide range of property thinkers, this work will focus on the ideas of the Progressive-era legal scholar Morris Cohen.[9] Drawing on the ideas of such contemporaries as Wesley Hohfeld and British socialist R. H. Tawney, Cohen produced a large number of works espousing a progressive, reformist approach to law and politics.[10] In particular, he expresses a view of ownership that downplays the Lockean emphasis on the individual and emphasizes instead the existence or absence of substantial collective benefit in evaluating a system of property. In his landmark essay "Property and Sovereignty," Cohen (1967) demonstrates the tight connection between the traditionally distinct realms of private and public power. Control over things (private power) via ownership, he argues, leads inexorably to control over people (public power). In the industrial capitalist world of the early twentieth century, for example, Cohen (1967) notes that property owners exert such tremendous control over those who lack property as to effect a kind of sovereignty over their lives.

This observation is not, in and of itself, an argument against private property. Instead, Cohen asks whether the sovereignty created by property rights in early twentieth-century America is preferable to any other form of governance. His answer is that property has a useful function but must be limited in its power in order to continue serving the common good. What kinds of limitations are desirable? Cohen (1967) promotes typically Progressive restrictions on private property, including a weakened power of bequest for individuals. He also includes government regulations requiring higher wages and better working conditions without compensating business owners based on a property right to specific regulatory expectations. More important, however, is the basic assumption of the intellectual exercise in the first place. Simply by viewing ownership as a malleable institution, Cohen takes the instrumentalist approach.

Despite their clear differences, Cohen's position is not a complete rejection of the Lockean system. In fact, he defends the idea of natural rights and natural law in general (1959). Nor is Cohen entirely hostile to the labor theory of ownership propounded by Locke and his followers. "[T]he labor theory contains too much substantial truth," he writes, "to be brushed aside" (1967, 52). The crucial difference between the two is that Cohen is skeptical of labor as a unilateral, prepolitical, or inviolate justification for ownership. For Cohen, as for Tawney and other instrumentalists, the labor theory of ownership is not in any way natural; it is justified only insofar as it serves greater social ends, such as encouraging hard work. Determining property's success at achieving other collective goals of society remains the primary method of analysis. A property right is not a trump against the greater needs of society; it is simply an instrument of public will, subject to manipulation by government to serve the greater good.

The Egalitarian View: P. J. Proudhon

The egalitarian view is the opposite of the secure possession arrangement described in Hume's theory. In its strongest form, the egalitarian argument supports property rights that must be allocated equally to every individual. Ongoing redistribution is required to maintain this commitment to equal shares for all, both at the moment of initial allocation and subsequently. Because the commitment to equality of distribution is paramount, the type of right is relatively insecure and yet still prepolitical, exempt from government meddling in pursuit of other goals.

The work of the nineteenth-century French philosopher Pierre-Joseph Proudhon is emblematic of this egalitarian view. Concluding famously that the modern version of private property is simply theft, Proudhon argues against any justification for ownership that is nonegalitarian. Persons would never agree voluntarily, he contends (in direct conflict with Hume), to any system of private ownership that left them with less than an equal share (Proudhon 1994). This does not mean, however, that Proudhon rejects any form of ownership at all, as at least one modern philosopher inspired by the French thinker has confirmed (Christman 1994). Rather, there can be no ethical basis for allocation other than an equal distribution based on our shared existence as persons (Proudhon 1994). This equal right to property is clearly prepolitical in the sense that any legitimate government would have to respect it. The self-evidence of this need for equality permeates

Proudhon's writing: "What a profound disgust fills my soul while discussing such simple truths!" he concludes (1994, 66). "Can we doubt these things?"

In this manner, Proudhon explicitly rejects claims of ownership based on either labor or possession, as described by Locke and Hume, respectively. Both lead inevitably away from an equal distribution of wealth and therefore fail as justifications for private ownership (Proudhon 1994). His argument also rejects the adequacy of a Humean peace dividend to motivate or justify unequal shares of private wealth protected by the state. "... [P]roperty," he concludes (1994, 43), "to be just and possible, must have equality for its condition."

Although ownership based on equality is a prepolitical right, it appears in a much more insecure form than in the intrinsic account. Because property must be allocated on an equal per capita basis, and because the human population is always in flux, these rights must be weak and subject to regular redistribution. Thus Proudhon defends a usufructuary vision of ownership, in which persons have a use right to their property but few other powers (1994). Under this conception, we cannot destroy or transform our property, for example, and when we die, there is no private power of bequest. The propensity for redistribution is reminiscent of the instrumentalist account, but there is a crucial difference: redistribution is permitted in this case only to meet the continued goal of per capita equality, not to meet other various and changing social needs.

Taken as a group, the four perspectives provide a useful structure for classifying theories of ownership. Each perspective emphasizes different ideas. The Humean approach worries about creating secure property rights and grants them based on possession out of political necessity and due to the endowment effect. The intrinsic approach focuses on protecting the individual with a strong prepolitical right that must be respected by government at nearly all costs. The instrumental view emphasizes a very different justification for ownership, meeting the changing goals of society through an adaptable property structure. And the egalitarian ideal stresses the prepolitical right of every person to an equal share of property, based on our common humanity. Of course, few writers argue for a purely possessory, intrinsic, instrumentalist, or egalitarian approach to property. Nevertheless, these archetypes allow us to describe different approaches to property as tending more to one or another of these positions.

The various answers to the type and justification questions for property rights can also be summarized in a pair of typologies. In the first typology (Figure 2.1) the theories vary from a secure to an insecure property right along the horizontal axis, and from a prepolitical to a political justification for ownership along the vertical. As an intrinsic rights theorist emphasizing a secure power of ownership, Locke is located in the lower left-hand quadrant. Cohen, an instrumentalist seeking a weaker form of ownership, is at the opposite end of the spectrum. Hume lies in the quadrant supporting a strong right of property on a political basis, and Proudhon lies directly opposite in supporting an insecure, prepolitical right.

The second typology (Figure 2.2) arranges the theories according to their justifications for ownership. Here the degree of redistribution permitted is the x-axis variable, and the emphasis on individual versus collective goals is on the abscissa. Locke is again in the lower left corner, representing an indi-

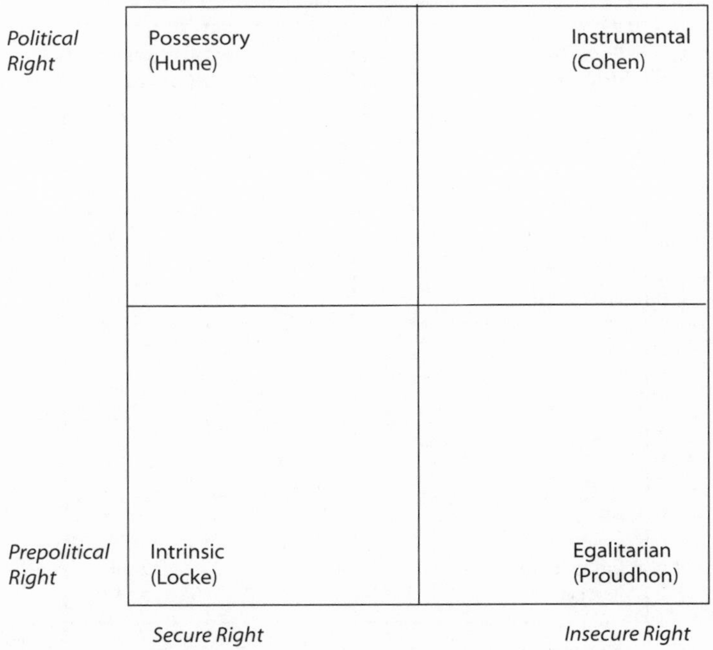

Political Right	Possessory (Hume)	Instrumental (Cohen)
Prepolitical Right	Intrinsic (Locke)	Egalitarian (Proudhon)
	Secure Right	*Insecure Right*

Figure 2.1 A Typology of Property Theorists—Type of Right

vidual-focused justification that rejects redistribution. Hume is in the upper left, presenting a nonredistributive theory on the basis of its wider social benefits. Cohen is in the upper right again, with a theory based on collective needs but permitting a great deal of redistribution. Finally, Proudhon's theory is once more at lower right, representing both Locke's emphasis on the individual and Cohen's willingness to redistribute.

The typologies in Figures 2.1 and 2.2 provide an initial basis for examining equity-based arguments about property and serve as the foundation for the theoretical framework applied in Chapters 3, 4, and 5. However, they are only the first steps toward that goal. Each of the property theories considered in this section gives rise to a family of specific allocation rules by which property rights, including licensed property, are appropriately distributed. The next section, therefore, will look at these allocation rules in more detail in light of the four theoretical perspectives described here. Matching concrete principles of distribution to their guiding property theories will help classify the allocation rules that were proposed, accepted, and rejected in the empirical cases that follow.

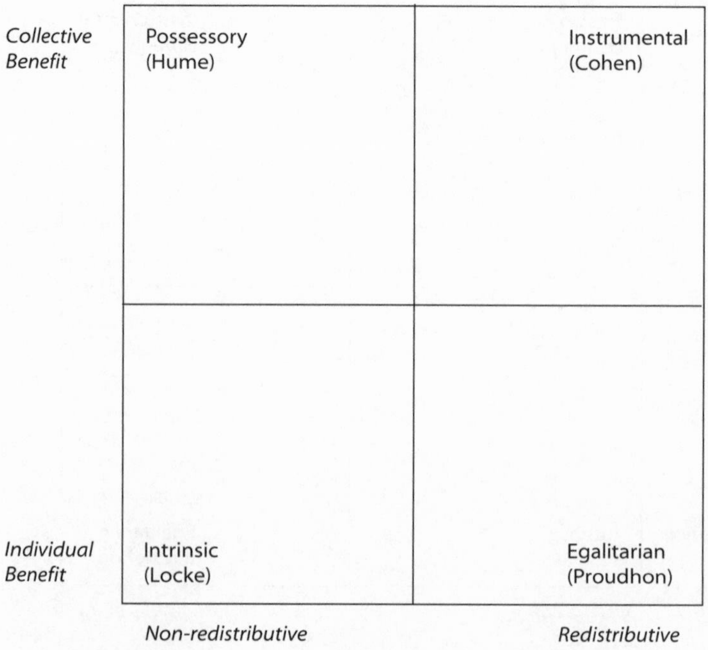

	Non-redistributive	*Redistributive*
Collective Benefit	Possessory (Hume)	Instrumental (Cohen)
Individual Benefit	Intrinsic (Locke)	Egalitarian (Proudhon)

Figure 2.2 A Typology of Property Theorists—Justification for Ownership

ALLOCATION RULES

Like a theory of property, a fully specified allocation describes both the type of right being created and the standards for distributing those rights. This section will consider some of the allocation rules indicated by the views of property just presented. Although actual allocations of licensed property rarely follow a single principle, they do draw extensively from the set of rules presented here. This section is therefore an intermediate step; only with the addition of the Hegelian perspective and others attempting to combine various principles will the property theory framework for the three case studies be complete.

Possessory Allocations: Grandfathering

Hume's theory of property stresses security for current resource holders. This goal indicates a need for rights that are well protected against confiscation or outside interference. It also suggests rules that allocate property to current possessors for practical reasons of political consent. An allocation of rights in this manner without cost to present users is commonly known today as grandfathering. Thus, air pollution permits might be grandfathered at no cost to current polluters according to their latest emissions figures. The vital criterion for grandfathering is possession: holding or occupying the resource at the time the allocation is made. How one gained possession of the resource or otherwise achieved control of its use is not important; this is an ahistorical approach to allocation. Therefore, any equitable defense of the grandfathering approach will be pragmatic and utilitarian, relying on the need for mutual advantage and the endowment effect, as Hume's argument does.

As a distributive strategy based on current possession, grandfathering should not be confused with allocations based on other specific actions. Intrinsic allocations based on labor, for example, do not distribute property according to some standard of Pareto improvement. They allocate based on the recognition of individual rights. Although the two allocation rules may indicate the same outcome in some cases, such overlap is neither necessary nor obvious. Those currently in possession of a resource may not be the same persons who mixed their sweat with it; rather, possession could result from a variety of means besides productive labor, including exchange, simple occupation, or theft. Where there is such a divergence, the possessory allocation will go to those who currently hold the resource, regardless of the previous expenditures of labor by their rival claimants.

This important distinction is often lost in the literature on market-based policies. Consider the following example: Two men claim a tract of land. One invested his labor on the parcel for several years, the other has recently succeeded at blocking the first from the land without actually putting it to use himself. The Humean system of grandfathering would award the tract to the current possessor—despite his lack of prior use—but the Lockean system would grant ownership to the land user now estranged from the property. Nor is this a merely theoretical distinction: as we shall see, allocations in the acid rain and grazing cases both favored productive labor over mere possession or occupation in just this way, whereas arguments in the climate change negotiations favored the possession standard instead. Thus, the two theories generate different allocation rules that do not always point to the same conclusion, and it is the Humean system with its emphasis on current possession and Pareto improvement that best captures the modern idea of grandfathering.

Intrinsic Allocations

Allocations based on the intrinsic view of property use the Lockean idea of sweat equity, or what I will call prior use, as the criterion for distribution. The type of right allocated is typically quite secure in accordance with Locke's theory. Unlike grandfathering, the intrinsic allocation strategy is fundamentally historical in nature, looking at patterns of prior resource use for guidance in allocating licensed property rights. Indeed, the prepolitical nature of this vision of property makes the phrase *allocation of rights* misleading in this case. In reality, an allocation based on an intrinsic view of ownership recognizes existing, prepolitical property rights within new legal structures, rather than creating them in some *de novo* process. Intrinsic allocation rules are prevalent in the United States and played a leading role in the acid rain and grazing cases.

Allocations of this type frequently begin with a "baseline"—a historical level of resource use by an individual or entity at a specific time prior to the implementation of the licensed property system. A simple baseline in an individual transferable quota program, for example, might be the total catch of a fishing boat in the previous season. Baselines can be determined on the basis of a single year or by averaging several years of prior use. Determining the precise method of setting the baseline is itself a thorny policy problem, given the possible variants and their potentially significant distributive impacts. Although the specifics may vary, the baseline must rely on some measurement of actual resource use (and not simply possession) to be a historical allocation rooted in Lockean theory.

In addition, different kinds of prior use carry different weight under the intrinsic approach. Resource use that approximates the productive personal labor favored by Locke is most likely to be rewarded with ownership rights. In particular, an intrinsic allocation recognizes prior uses that are *tangible* and *beneficial*. A tangible use means one with lasting and easily observed effects on the resource in question. Fencing, plowing, and harvesting crops from a plot of land would be a prime example of a very tangible use. Similarly, the more a prior use is perceived as beneficial or productive, the more likely it is to be rewarded in an intrinsic allocation. As the direct production of food for human consumption, farming is also a strong example of a beneficial use.

Prior use that is more commonly perceived as something other than productive labor, such as an externality of an industrial process like air pollution, is less likely to gain consideration for allocation on intrinsic grounds. Air pollution is less tangible, in that emissions are often invisible or at least leave no obvious lasting trace. And it is also less directly beneficial, since it represents the undesired and negative outcomes of industrial processes. Indeed, air pollution is much closer to mere possession or occupation of the resource (the atmosphere) as described in the Humean account above. Thus, to the degree prior uses are more closely aligned with the Lockean ideal of beneficial and tangible personal labor, intrinsic allocation rules are more likely to gain recognition and influence. All other things being equal, in other words, one would expect prior use claims based on farming and ranching to be more politically effective than those based on industrial air pollution.

Instrumental Allocations

Instrumental rules reject the strict and unchanging allocation criteria of the intrinsic and possessory views. Following Cohen, the needs of society are paramount—needs that can and do change over time. Therefore, the right created is less secure and the rules for allocation can considerably modify historical or status quo use patterns. Although the number of possible instrumental allocation strategies is virtually endless, they can be grouped into two general categories: share-based and class-based methods. In practice, both share- and class-based rules have a large impact on licensed property allocations, especially in the domestic U.S. context, as they clash with intrinsic-view principles during the policymaking process.

Share-based allocations distribute property rights according to rules that limit the holdings of any particular owner. For example, an initial distribution of equal shares of property to all users of a resource is a share-based allocation scheme. (Note that this is not the same thing as a prepolitical right to an equal share of property á la Proudhon, which is based on individual rights and must be maintained in perpetuity.) Other share-based allocations tolerate more inequality, as can be illustrated by two examples from public lands grazing. A maximum limit on shares, found in early Forest Service grazing regulations, caps allocations for all owners at some specific level regardless of other factors. A minimum or protective limit, also evident in both Forest Service and Grazing Service regulations, guarantees a certain amount of property to all users, again regardless of other considerations.

In contrast, *class-based* rules distribute property rights by privileging the claims of some resource users over others. Rather than restricting shares, this type of allocation restricts eligibility to receive any property at all. A class-based rule, for example, might limit allocation to citizens (as occurred in the grazing case) or to the owners of private capital used in resource exploitation rather than hired workers (as is done, controversially, in many fishery quota allocations). Class-based rules also function according to equity considerations but in the service of different goals than their share-based peers. Qualities of prospective owners, rather than quantities to be owned, are the guiding criteria.

Allocation via *auction* is a third instrumental alternative. An auction allocates resources to those most willing to pay, rather than to current possessors, thereby maximizing aggregate social welfare created by the distribution. It strays from the principles of Hume's theory by rejecting Pareto improvement as important and differs from Locke's account by ignoring historical claims of entitlement. Strongly redistributive and emphasizing collective benefits, auctions are the ultimate instrumental allocation mechanism. It is worth noting that despite their enthusiastic academic supporters, auctions have played a more limited role in many allocations of licensed property rights to date, particularly in contexts where prior use is already established. This restricted role is consistent with the tendency of allocations in the three case studies to rely on a synthesis of opposing perspectives, rather than a unique allocation principle.

Many allocations based on intrinsic or Humean principles might also appear to qualify as class-based. An allocation based on prior use, for example, might be described as an allocation to the class of those with a history of use. Or an allocation based on Pareto improvement might be described as rewarding the class of current property possessors. Calling these allocations instrumental is incorrect, however, because instrumental rules must be mutable and potentially redistributive over time, and are political rather than prepolitical in nature. Subject to present and future adjustment, they exclude a Humean allocation based on a desire for greater security. And as political rules of entitlement, they allow further government intervention in pursuit of broad social goals in ways that intrinsic rules based on prior use do not.

Egalitarian Allocations

Allocations faithful to Proudhon's view of ownership are quite simple: they are equal per capita distributions. Factors that are critical in the other allocations, like historical use, current possession, or even social needs, are irrelevant here. Equal shares for all is the rule regardless. In addition, the type of right to be allocated must be relatively insecure, usufructuary, and subject to continued redistribution to preserve per capita equality over time. Such rights cannot otherwise be tampered with by political processes, however, since the commitment to equality is paramount. Although egalitarian allocation rules do not appear in the U.S. cases studied here, they are important in the international context of climate change. Indeed, there are some striking theoretical links between Proudhon's ideas and various proposals for equal per capita allocations of atmospheric capacity for greenhouse gases (e.g., Agarwal and Narain 1991). This connection will be discussed in more detail in Chapter 5.

Using the allocation principles just summarized, one can describe the many equity arguments present in the licensed property cases covered by this work. Because arguments about policy tend to revolve around specific allocation rules rather than the theories of property that support them, establishing the connection between the two is critical. Few people talk explicitly about Lockean rights to grazing permits, but many advocate allocation rules on equity principles that make sense only from a Lockean perspective. These theories and the specific allocations they embrace are summarized in Table 2.1, at the end of this chapter.

In the real world, of course, few allocation strategies simply adopt the approach of only one theorist. In particular, rules based on both intrinsic and instrumental ideas are prevalent in the U.S. policies considered here, whereas a conflict between possessory and egalitarian entitlements seems to be building in the climate change case. The result in all three examples is an allocation debate featuring two competing and contradictory approaches. Given this recurring dilemma, alternative visions of property that try to resolve the conflict between diametrically opposed property views would be useful. Fortunately, Hegelian property theory provides at least one such alternative, focused on the tension between intrinsic and instrumental property ideas so controlling in the acid rain and grazing cases.

HEGELIAN PROPERTY THEORY: A "FRACTIOUS HOLISM"

Hegel provides a view of property that deliberately embraces the opposition between rights and duties framed by Cohen and Locke. He seeks, according to one interpreter, a "third way" between the opposite extremes of "anarchic individualism" and the "repressive collectivism of unopposed state power" (Brod 1992, 8). His view accepts the individual right of ownership while trying to harmonize it with socially oriented controls. In the end, according to Hegel, society will synthesize these two opposing roles for property in a dialectical manner. Given that the allocation outcomes described in the acid rain and grazing cases are based on a reconciliation of intrinsic and instrumental principles, his ideas are critical to this study.

This section presents Hegel's theory of property and the allocation rules it suggests. Like the theory that inspires it, a Hegelian allocation is a complex combination of intrinsic and instrumental ideas. It seeks rules that recognize the intrinsic entitlement to ownership while modifying it in a selective manner for the collective good. This merging of allocation rules is evocative of Bennett's idea of a "fractious holism," mentioned in the

Introduction, which will be revisited in more detail here. A Hegelian allocation also describes not only a specific allocation outcome but also the process for creating it. That process is dialectical: starting with intrinsic rules, the proposed allocation moves toward the opposite instrumental position before settling on a system of rules that synthesize the conflicting ideas. This Hegelian process is remarkably similar to the path followed by policymakers in both domestic cases and offers some suggestive parallels for the distinctive and yet related conflict unfolding in the international context regarding climate change.

Hegel's Dialectical Theory of Ownership

Ownership is critical to Hegel's conception of human development in a way that is unlike many, if not all, of the political philosophers who preceded him. He begins his analysis of political philosophy in the *Philosophy of Right* (1967), first published in 1821, by viewing humanity as beings with free will. But despite this basic freedom, they initially lack any means of connecting their free will to the external world that surrounds them (1967, sec. 41). It is property that first provides this connection. Hegel contrasts the freedom of human will with the world of "things"—objects external to humans that are not free and therefore, in Hegel's view, are without rights (sec. 42).[11] A property right is created by an individual's freely made decision to claim ownership of the external object (sec. 44). Ownership can be communicated through physical possession accompanied by clear intent to own, or simply by marking the object in a manner recognizable as property by others (sec. 54).[12] Mere possession is inadequate; the expression of personal free will to own the object is required (sec. 45). In this manner, humans possess a prepolitical right to ownership by unilateral action that fits under the intrinsic right idea. This is the initial view of property in society.

The philosophical primacy of property makes Hegel's account quite different from those of his predecessors. Even though both his account and Locke's stress the intrinsic nature of the property right, under the Hegelian view, ideas as fundamental as a right to life or liberty, for example, are grounded in the notion of "self-ownership." The active occupation of one's own body by one's free will is what prevents control or ownership by others (Brod 1992). Property therefore is not simply a means to create wealth or to distribute goods appropriately (as it might be in the Humean or the instrumentalist accounts), nor is it a right that simply protects the individual from the state (as in the intrinsic account). It is instead a fundamental articulation of our freedom and our humanity—the first expression of our free will that gives us a personality and a set of rights distinct from the rest of the natural world.[13]

While recognizing property's initial focus on the individual, Hegel's account of private ownership takes notice of its significant social aspects as well. This inclusion again contrasts him with his predecessors, such as Locke, who emphasized the gains to the individual under a private property regime. For Hegel, property is the very basis of interpersonal relations—the initial means by which two free individuals learn to recognize and interact with each other. Only ownership of an object by one person's free decision prevents a claim of ownership by another (Hegel 1967, sec. 50). Because of this, a property claim must be recognizable by others before it becomes legitimate: "my inward idea and will that something is to be mine is not enough" (sec. 51).[14] These social aspects of ownership serve to modify, but not replace, the initial focus of property on the rights of the individual.

The importance of "social consensus," as Shlomo Avineri (1972, 89) describes it, to maintaining a healthy system of property is therefore what makes Hegel's view complex. Unlike an intrinsic theorist, Hegel is deeply concerned about the need for restraining the individual power of ownership in some manner. He rejects any theory, such as Locke's, that "makes the securing of property rights the principal basis or task of the state" (Brod 1992, 65). Like an instrumentalist, Hegel seeks other institutions, such as the family and the state, to serve as a balance to the individual power of ownership. Hegelian property rights are not to be respected by society at all costs. They are subject to modification and constraint according to competing social needs (Ryan 1984; Stillman 1980).

Hegel limits the modification of property rights, however, in a manner wholly foreign to the instrumentalist viewpoint. The individual autonomy created by private property is itself a collective good that may not be subsumed under other social goals (Brudner 1995). In many circumstances, the Hegelian policymaker must therefore refrain from modifying individual property rights even though the temptation to do so in pursuit of other collective goals is strong. To be permitted, the proposed modification must remain consistent with the benefits of the intrinsic private right (Brudner

1995). Otherwise, the initial intrinsic approach to ownership would be nullified rather than supported by the instrumental rule. Within the Hegelian synthesis both opposing principles must remain viable: neither is proven wrong, yet both are revealed as incomplete. Thus, as Avineri notes (1972), the Hegelian state is a constant threat to individual rights of private property, even as it insists on the continued legitimacy of such rights. The tension never fully dissipates.

Although it incorporates both intrinsic and instrumental ideas about property, the Hegelian view should not to be confused with a pluralist theory of ownership. A pluralist tries to pick and choose between intrinsic and instrumental principles (among others) according to some metacalculus of their relative degree of importance (e.g., Munzer 1990). Although similar in its respect for more than one side in the conflict, the Hegelian theory attempts to synthesize two opposing perspectives in a dialectical manner rather than provide a method of choosing between them on a case-by-case basis. Many allocation alternatives acceptable to a pluralist would fail a Hegelian standard, as the next section will show. The Hegelian view is neither a compromise nor an ambiguous waffling between opposing theories—it is an attempt to reconcile them.

To summarize, for Hegel private property ownership remains crucial to human development in a manner that is fundamentally different from a mechanism of individual wealth maximization. Property develops and relies on important individual and collective goals as an institution. It begins by serving individuals but then recognizes a tendency toward atomism and selfish behavior that must be checked. Yet property is not simply a mechanism for furthering social goals, as it is for an instrumentalist. It is crucial to the individual's most fundamental relationships to the world around her. Within the Hegelian account, the idea of property honors and includes the individual roots of ownership as it provides for limited social constraints upon it. Therefore, Hegel's theory of property is uniquely suited to the conflicting notions of property within certain licensed property policies.

Equity-Based Allocations: The Hegelian View

Given Hegel's dialectical approach to ownership, what would a Hegelian allocation of rights look like? Clearly, the outcome would have to include both intrinsic and instrumental rules, while seeking some sort of coherence between the two in accordance with Bennett's idea of a "fractious holism." In terms of process, a Hegelian allocation would start from an intrinsic perspective, subsequently moving toward an instrumental view and trying to reconcile the two. Thus, an allocation must meet both outcome and process standards to be defined as Hegelian.

Because it is the key to understanding a Hegelian allocation of property, it is time to consider more clearly what *fractious holism* means. Bennett does not apply the idea to property theory herself, but her explanation with respect to modern environmental conflicts is analogous in many ways. Bennett (1987) believes that environmental policy has reached a stalemate because of the entrenched opposition of two perspectives, which she calls environmental management and natural holism. Environmental management seeks to impose complete human control over nature; natural holism seeks to restore humanity's supposedly cooperative, symbiotic place in the natural world. Each argument relies on the other to define its own position and to look more reasonable by contrast. But in reality, as Bennett (1987) points out, both share a similar but unstated perspective on the world—that nature is somehow able to be completely in harmony with human interests.

Bennett rejects this conclusion and argues instead for a fractious holism of the conflicting views rooted in Hegelian thought. In this example, *fractious holism* means an ontology of the world not fully subject to control by or harmonization with human interests (Bennett 1987). This view is holistic in recognizing the vital connections between humans and nature, but fractious because it concludes the interests of the two are never fully in harmony. Nature, in this account, is always partly an adversary or unknown "Other": something outside full human understanding. Fractious holism urges us to recognize this opposition while trying to work with nature as best we can, rather than trying to master it or pursuing some mythical notion of perfect harmony. More generally, fractious holism echoes Hegel by requiring that we recognize both the connections and the irreconcilable differences between two opposing concepts. It admits that there is a negative quality to many "goods" in life, that there is "chauvinism in patriotism, authoritarianism in community, neurosis in family life, and jealousy and rage in love" (Bennett 1987, 158). Thus, to be a fractious holist, if you will, is to be both honest and reflective about these dualities and the ongoing difficulty of reconciling them, and it is this aspect of the concept that is especially relevant to conflict over property.

Intrinsic and instrumental theories of ownership are prime candidates for a fractious holism perspective. Supporters of the two ideals categorically reject their opposing arguments with great frequency, resulting in policy stalemates not unlike the difficulties in environmentalism described by Bennett. Each theory relies in part on this rejection of the opposite view to define its own position, and both share a fundamental underlying belief: property need not be allocated according to an initial social consensus. Both have dualities or dark sides that require acknowledgement by political actors: intrinsic rights threaten radical inequalities of wealth, and instrumental rights threaten political tyranny. Thus, decisionmakers faced with a conflict between intrinsic and instrumental arguments are frequently forced, in Bennett's (1987, 158) words, "to confess that we lose something by acknowledging the underside of our ideals: we lose the ease of our conscience and the freedom to act without tortured consideration of possible implications, long-term effects, dangerous unintended consequences."

Given the details of this helpful concept, what would a Hegelian allocation in the fractious holism tradition look like? Most strongly, a Hegelian allocation would start with prior use as the first proposed standard for assigning rights. Unlike a purely intrinsic approach, however, the Hegelian technique would then modify that prior use standard in a way that respected and was consistent with the original goals of the intrinsic rule. For example, a system that rewarded prior use to the degree it was beneficial but modified it instrumentally to the degree it was harmful would qualify as Hegelian. Such a rule might provide property rights up to a certain level based on prior use but decline to offer previous users any additional allowances that would cause ecological damage. Or the allocation might use a class-based rule to reward certain kinds of prior use over others as being more beneficial to society. The crucial question for any Hegelian allocation is whether and to what degree the instrumental rules preserve the integrity of the intrinsic principles that they seek to modify.

In practice, a Hegelian allocation might be easily confused with any compromise between allocation principles based on intrinsic and instrumental approaches. Although the Hegelian allocation resembles a compromise in some ways, it has several additional qualities, described in Box 2.1. A Hegelian allocation specifically starts from the intrinsic perspective and then moves in the instrumental direction; an allocation process that began with a share-based rule and then modified it over time by prior use

Box 2.1 Qualities of a Hegelian Allocation

1. *Process:* starts from the intrinsic position.
2. *Process:* moves toward the instrumental position over time.
3. *Outcome:* attains a fractious holism between the two principles that is
    a. *coherent:* not simply an inconsistent or nonsensical pastiche of rules, and
    b. *dialectical:* does not destroy the normative authority of the intrinsic rules rewarding beneficial prior use even as it adds instrumental modifications.

would not be Hegelian. Furthermore, the allocation must attain a fractious holism between the two opposing theories. This quality is harder to define, but it certainly excludes a number of possible compromise outcomes. Allocations that inconsistently use specific intrinsic or instrumental rules, for example, are not Hegelian. Those that fail to retain some significant degree of respect for the initial set of intrinsic principles would also fall outside the Hegelian model. Thus, only an allocation that (a) starts from the intrinsic perspective and then (b) moves in an instrumental direction toward a (c) coherent and dialectical set of intrinsic and instrumental rules meets the Hegelian standard.

As noted earlier, most real-world allocations of licensed property fail to follow exclusively an instrumental, intrinsic, or any other single approach. Review of the two U.S. cases studied here will show that their fit with a Hegelian approach is much better. Neither the Clean Air Act nor the Taylor Grazing Act cases were perfect examples of Hegelian allocation processes or outcomes in action, but they were much closer to the Hegelian perspective than to any other single theory of property. The Hegelian vision of ownership is thus a useful guide to both past and future allocation problems where there is a tension between the intrinsic and the instrumental perspectives.

IV. BEYOND HEGEL: OTHER EXAMPLES OF FRACTIOUS HOLISM?

Allocations in the climate change case remain too incomplete to establish whether they will eventually fit with the Hegelian model. The initial indications, however, are that an entirely different debate is occurring. Other property theories, including possessory and egalitarian arguments, are more influential in this particular allocation, as will be reviewed in some detail in Chapter 5. This difference makes the Hegelian model, with its synthesis of intrinsic and instrumental principles, less obviously relevant. However, the tension and conflict between the Humean and egalitarian positions in the climate change process resemble in many ways the polarized conflicts between intrinsic and instrumental principles in the acid rain and grazing cases. Thus, another distinctive but still unspecified fractious holism between property rights based on possession and those based on strict equality may have a parallel importance in this case, as well as others with similar institutional settings and patterns of prior resource use.

One possible source of such reconciliation could be the property theory of the eighteenth-century German philosopher Immanuel Kant, particularly as he has been summarized and interpreted by several modern commentators. Like Hegel, Kant seems to be recognizing two conflicting ideas of ownership simultaneously in his approach to property. As Alan Ryan (1987, 79) puts it, Kant asserts both that people are valid owners only "when a legal order gives them that property" and that we have "a 'natural right' to appropriate unowned things and make them our property." In this manner, Kantian property theory combines the political and prepolitical perspectives on ownership in a manner that so far sounds something like the Hegelian view.

In Kant's theory, however, the primary objective of a property system is securing one's possessions from others, as in the Humean account. The role of the state is to protect and secure these possessory rights as property and prevent their loss through coercion or intimidation by others (Ryan 1987). Kant argues that individuals therefore have a right to create a civil society based on unequal property relations, including the right to coerce others into participating in this property system (Williams 1977). In accepting this right of coercion, Kant goes a step beyond Hume's notion of universal consent to property based on Pareto improvement. His approach recognizes that some individuals will be losers under the transition to a property system, but that at least under some conditions this is an acceptable outcome.

Yet Kant's conception of property is not simply a guarantee of possession and security. In fact, it is also based on a second contradictory idea: the common possession of the earth's resources in a theoretical "state of nature" (Williams 1977, 35–37). As another interpreter has described this principle, "common possession is [Kant's] point of departure for all private ownership. ... Private ownership must respect the prior moral agency and common ownership of all" (Gillroy 2000, 215). Kant does not necessarily argue that some state of nature, with ownership in common for all, actually existed in human history. Rather, it is "a logical fiction" necessary for an ethical defense of unequal property rights based on possession in civil society (Ryan 1987, 80). In other words, only when we can justify coercive and unequal distributions of property with respect to some logically prior notion of property for all will we have created a truly ethical system of ownership.

This additional principle of common possession in a state of nature is highly reminiscent of Proudhon's natural right to an equal share of the world's resources. In Kant's theory of property, therefore, an egalitarian notion of common ownership appears to serve as a fundamental correction to Humean principles based on security and possession. This tension leaves one with another fractious theoretical outcome, if not one that is entirely holistic. As Williams (1977, 40) puts it, "We have a situation, then, where the whole body of citizens is seen as safeguarding an institution which it has in the first place to be coerced into accepting." As in the Hegelian account, the tension between the two conflicting perspectives remains ongoing. However, Kant offers us less guidance on how to resolve this tension in an allocation context. It is, as Williams (1977, 40) concludes, "a lack of coherence in Kant's argument" that may simply point out "a difficulty which lies at the heart of Western society."

That said, more than one modern philosopher is working on a theory of property rooted in the egalitarian tradition of Proudhon while acknowledging the importance of more Humean ideas, like security of personal possessions and protecting individual autonomy (Gillroy 2000; Christman 1994). One modern interpreter in particular offers an innovative principle for reconciling the two perspectives: an allocation with a guaranteed minimum of property for all. John Martin Gillroy (2000) has recently applied Kant's political philosophy to environmental policy at great length. As part of this larger project, Gillroy offers a specific interpretation of how Kant's property theory would approach a distribution of ownership rights. His view is that the Kantian approach requires a baseline of property for everyone, even while it permits continued inequalities beyond that minimal level (Gillroy 2000, 241). Kant's political system requires an absolute equality of freedom, or autonomy, for all individuals, and such freedom necessitates at least a minimum amount of material possessions for everyone. Although Kant himself appears not to have directly espoused such a universal right to property (Ryan 1987), according to Gillroy (2000), his theory implies this basic right of material ownership.

Extending that particular view of Kant to the issue of licensed property allocation, one gets something resembling a protective limit guaranteed to all. Unlike Proudhon, Kant's approach clearly tolerates a significant range of inequality among property owners in the interest of security and respect for possessory rights. Unlike Hume, however, Kant's theory may also imply a basic level of allocation for all parties, regardless of the redistributive implications. This may not exactly be a fractious holism between the possessory and egalitarian positions, but it clearly seems to parallel the Hegelian approach in its attempt to reconcile two diametrically opposed property perspectives. Thus, Gillroy's interpretation offers a promising resolution to the tension in Kantian property theory, making it a leading candidate for resolving conflicts between Hume and Proudhon, as found in the climate change case.

Alternatively, the ultimate lack of coherence in the Kantian account of property also suggests a more pessimistic conclusion: that the tension between Humean and egalitarian principles found in the climate change case may be too difficult to resolve on an equitable basis. If this is in fact the case, one might expect to see ongoing polarization and conflict between entrenched normative views in this setting. Although the climate change process remains incomplete, there is no question that conflict over allocation has only deepened and intensified since agreement on the Kyoto Protocol in December 1997. In this respect, the lack of a prominent normative perspective on property that can reconcile these particular views may be part of the difficulty. This possibility will also be explored further in Chapter 5.

SUMMARY

This chapter placed theories of property and the allocations they support into a framework for analyzing the empirical cases of Chapters 3, 4, and 5. It reviewed four prominent theories from the literature on property and explained how they differed in terms of the type of right assigned and the justification for ownership supported by each. It then turned to the work of Hegel and Kant to seek theories that might help reconcile specific conflicts between these principles along the intrinsic-instrumental and the possessory-egalitarian axes, respectively. In the case of Hegel, this reconciliation seems well specified and fully applicable to the acid rain and grazing cases. In the case of Kant, however, the reconciliation remains a work in progress, both theoretically and in terms of the allocation process for greenhouse gas emissions, considered in Chapter 5. Whether Kantian principles can effectively combine the Humean and egalitarian positions, or whether *any* property theory can do so effectively, remains to be seen.

The five views of ownership (excluding the still indeterminate Kantian perspective) discussed in detail and their associated allocations are summarized for convenience in Table 2.1. The table presents each theory in a separate column and lists the specifications of an allocation strategy for licensed property associated with that property idea, as well as the justifications for ownership, the type of rights assigned, and the guiding theorist. This table is one way of presenting the full theoretical framework needed to analyze the empirical allocations in the chapters that follow.

Finally, it is now possible to revisit the two-by-two typology of property theory in a useful manner. Figure 2.3 combines the type of right and justification variables from Figures 2.1 and 2.2. In addition, Hegel appears in the grid in a unique position that reflects his attempt to reconcile the conflicting intrinsic and instrumental positions. This typology will be the framework within which one can locate the allocations and arguments found in the three empirical cases.

Table 2.1 Property Theories and Allocation Strategies

Property concept	*Possessory*	*Intrinsic*	*Instrumental*	*Egalitarian*	*Hegelian*
Justification for ownership	Possession; Pareto improvement	Labor; prior use	Changing ideas of social good	Common humanity	Individual action with social recognition
Type of rights assigned	Political; secure	Prepolitical; secure	Political; insecure	Prepolitical; insecure	Dialectical; varies
Recommended allocation rules	Grand-fathering; possession	Prior use; sweat equity	Share- and class-based rules; auctions	Equal shares for all	"Fractious holism" of intrinsic and instrumental
Theorist	David Hume	John Locke	Morris Cohen	P. J. Proudhon	G. W. Hegel

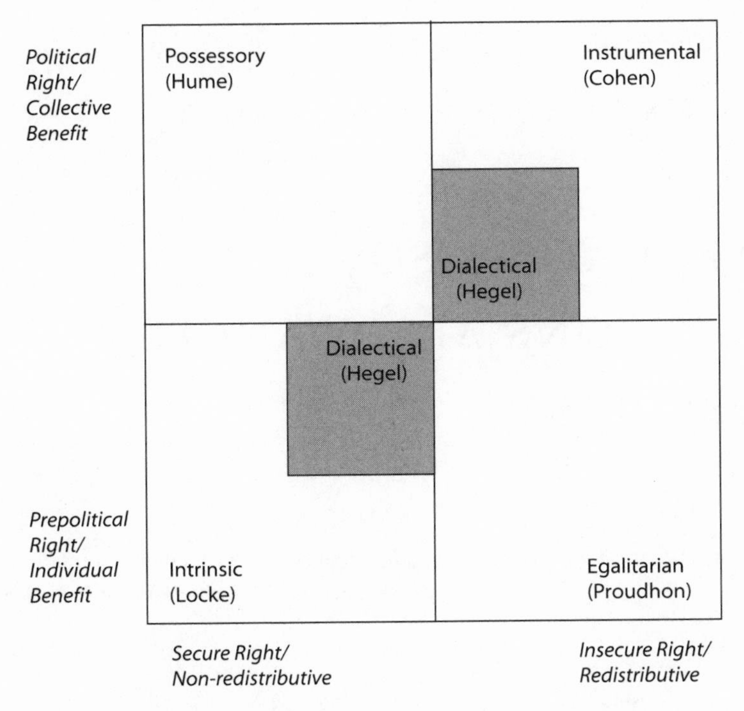

Figure 2.3 A Typology of Property Theorists—Type and Justification Combined

Now that the theoretical tools have been fully presented, it is time to start applying them to the practical world. The first case will be the most widely known example of a market-based environmental policy to date, Title IV of the Clean Air Act Amendments of 1990. Using the property framework just developed, the next chapter will consider the policymaking process behind the act in a new light, demonstrating the prominence of equity-based arguments in the process and showing the relevance of property theory to explaining them.

CHAPTER 3

ALLOCATING SO_2 EMISSIONS ALLOWANCES, 1989–1990

Allowance allocation is primarily a political process, not an environmental one.
— *Brian J. McLean, director, EPA Acid Rain Division*

Imagine a government program creating more than a billion dollars in legal assets overnight. Imagine further the government process of distributing that wealth. For many people, the images that come to mind are likely ones of unbridled avarice and greed. Certainly, from a rational choice perspective on political action, one would expect to see a prevalence of self-interested behavior by legislators seeking to maximize the portions allocated to their regions. Fairness, or equity, seems an unlikely factor in the process. At best, one might expect to see the rhetoric of fairness obscuring a real game of bare-knuckle politics in maximizing one's own share.

The Acid Rain Title of the 1990 Clean Air Act Amendments (CAAA), of course, was just such a program: creating and distributing new wealth in the form of emissions allowances. Yet self-interested behavior by legislators in favor of their regions has served as an incomplete explanation of the allocation process at best. In fact, norms of fairness and equity played a crucial role in determining the allocation by Congress. Furthermore, the content of the relevant norms was quite specific, grounded in the conflicting intrinsic and instrumental principles of property theory. The result was an allowance allocation that eventually melded these two conflicting normative positions into an outcome that is consistent with the Hegelian approach to ownership.

This chapter documents how those specific norms of equity shaped the allocation process for sulfur dioxide (SO_2) allowances under the 1990 CAAA. It argues that equity norms influenced the final allocation of allowances in several important ways. First, the participants in the acid rain debate in this period were outwardly obsessed with fairness and equity, drawing on rhetoric evocative of the intrinsic and instrumental positions described in Chapter 2. More significantly, the resulting allocation rules followed the rhetoric, struggling to combine intrinsic and instrumental principles regarding the ownership and allocation of emissions permits. In this manner, the primary axis of conflict in this allocation ran on one diagonal of the property theory framework presented in Chapter 2 and illustrated here in Figure 3.1. The importance of equity norms is further demonstrated

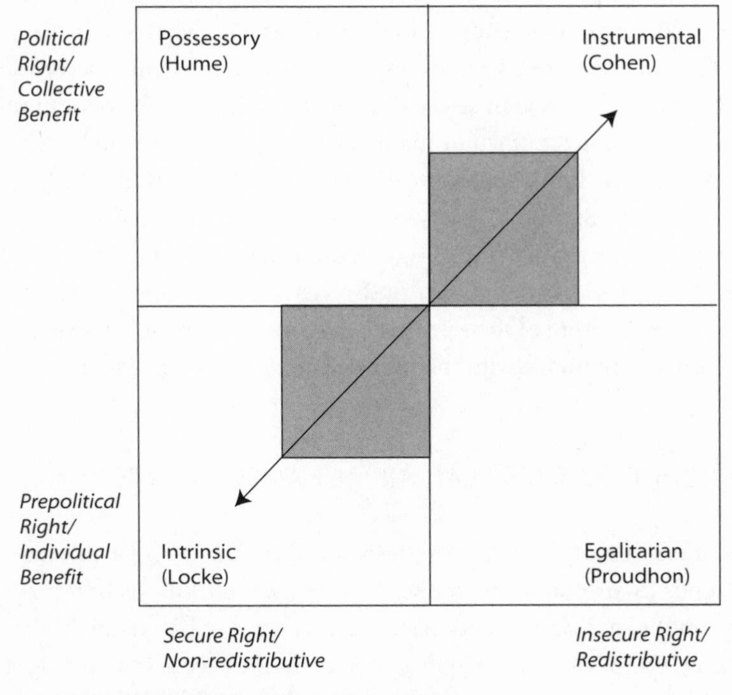

Figure 3.1 Axis of Primary Allocation Conflict: Acid Rain

by the fact that arguments following a consistent normative perspective in this process fared better than those that were normatively inconsistent.

In fact, both the allocation process and the final result in this case are substantially Hegelian, as defined in the typology from Chapter 2. Consistent with a Hegelian process, the bill started from a strongly intrinsic allocation and then incorporated certain carefully considered instrumental adjustments. But although intrinsic arguments had significant influence, their authority was reduced by the relatively negative and intangible nature of the prior use being defended—the emission of pollutants into the atmosphere. Arguments specifically in favor of historical emissions rates generally failed as a result, and the intrinsic perspective was less influential in this case than in the grazing rights allocation considered in the next chapter.

The chapter makes its argument in six sections. Section I presents a brief explanation of how the idea of marketable emissions permits, or allowances as they are termed in the law, came to be a viable policy option for the U.S. acid rain problem in the late 1980s. Sections II through IV document and discuss the various equity arguments appearing in the legislative process from 1989 to 1990, focusing on congressional hearings and debates over the allocation of allowances. These sections rely on data from the written record supplemented by interviews with selected participants. Because the allocation in the CAAA case was determined primarily by Congress, rather than by administrators at executive branch agencies, these three sections perform the bulk of the chapter's work. Section V briefly considers several equity issues related to the subsequent implementation and administration of the law. Finally, Section VI summarizes the chapter and reviews the rhetoric regarding the "unprecedented" nature of the acid rain program—rhetoric this research will at least partially contradict with the historical grazing case in Chapter 4.

BACKGROUND TO THE CLEAN AIR ACT AMENDMENTS

The acid rain title of the CAAA represented a confluence of two separate policy trends in air pollution control. The first was a cautious but growing interest among politicians and regulators in market-based methods for controlling pollution. The second was increasing concern during the 1980s, especially in the Northeast, regarding the need to reduce emissions of sulfur dioxide to control the problem of acid rain. To fully understand the genesis of the SO_2 trading program, one needs to understand each of these policy trends in more detail.

Market-Based Policies and the Clean Air Act

The Clean Air Act became law as part of a flurry of environmental legislation in 1970. Initially, the act concentrated on air pollution in urban areas. Automobile manufacturers were primary targets, as were industrial polluters and coal-fired power plants (Cohen 1995, 19–20). The law required the Environmental Protection Agency (EPA) to set limits, known as National Ambient Air Quality Standards (NAAQS), for concentrations of various pollutants in the air. The emphasis was on improving local air quality by reducing emissions in the immediate vicinity of large pollution sources.[1] All states then were required to describe how they would meet these NAAQS in a state implementation plan (SIP), subject to review and approval by EPA. Under the law, emissions standards were fixed and inflexible for all sources and were sometimes set at levels well beyond the capacity of current pollution control technology. Automakers, for example, were required to reduce vehicle emissions by 90% between 1970 and 1975—a goal that was unattainable with technology existing at the time of the act's passage. In short, the law was a clear example of traditional command-and-control regulation.

Perhaps not surprisingly, many of the standards and deadlines set by Congress were not met in a timely manner, leading to amendment of the law in 1977. Besides extending many deadlines for compliance, the 1977 amendments also reemphasized the command-and-control approach. All sources of pollution in "nonattainment" areas (those still failing to meet NAAQS) had to adopt new pollution control measures. New sources in polluted areas were singled out for the most stringent controls, being required to obtain the lowest achievable emissions rate possible. Existing sources in nonattainment areas only had to meet the less restrictive reasonably available control technology (RACT) standards. New sources in areas that were not yet polluted also had to meet stricter control standards by adopting the best available control technology (BACT). Frequently, the BACT standard required installation of expensive "scrubbing" equipment to remove pollutants chemically from the smokestack itself.

The Clean Air Act quantified many of its standards and rules in terms of a maximum permissible emissions rate for a pollution source. Arguments over emissions rates were a central part of the allocation debates in the 1990 amendments, so a brief explanation of how those rates are calculated is in order. Emissions sources like power plants both produce and consume energy. They produce electricity, which is measured in megawatts (MW). They consume fuel, such as coal, oil, or natural gas, to drive turbines that generate electrical power. This fuel consumption is often measured in British thermal units (Btus).[2] Thus, a hypothetical plant might burn 10 million Btus of coal to produce 1 MW of electricity. The quantities of energy consumed and produced are related to one another, but not in a fixed ratio; a more efficient plant, for example, might produce 1.1 MW from the same 10 million Btus of coal.[3] EPA then calculated SO_2 emissions as shown in Box 3.1.

The maximum statutory emissions rate for new sources of SO_2 in the original Clean Air Act was 1.2 pounds per million Btus, although the additional requirements of lowest achievable emissions rate or BACT forced most such plants to attain a rate of 0.5 pounds or lower. By comparison, older plants lacking modern pollution control equipment such as scrubbers frequently emitted SO_2 at rates of 5.0 pounds per million Btus or higher.

Despite continuing with a command-and-control approach, the 1977 amendments included several limited market-based mechanisms as part of the law. Prominent among these mechanisms were policies permitting offsets, netting, and so-called bubbles intended to facilitate more flexible and less costly compliance (Dudek and Palmisano 1988). Netting policies, for example, required sources in polluted areas to offset or "retire" existing sources of pollution before creating new ones. Bubbles permitted regulators

Box 3.1 SO_2 Emissions Formula

Fuel consumed (million Btus) × emissions rate (pounds per million Btus) = total emissions by weight

Units:

Fuel consumed: Millions of Btus
Emission rate: Pounds emitted per million Btus of fuel consumed
Total emissions: Pounds or tons emitted

Example:

A plant consuming 1,000 million Btus of fuel and producing emissions at a rate of 2.5 pounds per million Btus would produce 2,500 pounds (or 1.25 tons) of SO_2.

to treat multiple sources of pollution as a single entity, thus giving managers of those sources more freedom in meeting a single, aggregate emissions standard. As long as the emissions requirements for the bubble were met, individual sources within that imaginary enclosure could vary significantly in their specific pollution rates. Some bubble policies also included a provision for banking emissions reductions beyond the legal requirements for use in future years (Hahn 1989).

These market-based policies were used selectively throughout the 1980s, but regulatory uncertainty and misgivings among many environmentalists and EPA staff impeded more widespread adoption. The unwieldy union of limited market-based approaches to a dominant structure of command-and-control regulation also restricted their usefulness. For example, if a new plant still had to adopt the best technology available (via the BACT requirement), there was little room for market incentives to provide regulatory flexibility and lower the costs of compliance. Such impediments frustrated a small but vocal band of market-based policy advocates, two of whom invoked a racetrack metaphor by asking pointedly, "Why is this thoroughbred hobbled?" (Dudek and Palmisano 1988).

One counterexample to the relative ineffectiveness of market-based air pollution policies before 1990 was EPA's lead phaseout program (Hahn 1989), under which lead was slowly eliminated from all gasoline production. Starting in 1982, refineries were permitted to trade the "lead rights" newly required by EPA to include any amount of lead in gasoline. Such rights diminished over time for all companies, until they disappeared altogether and the trading program ended in 1987. In the five-year period of the program's existence, trading of lead rights was quite active: more than 50% of all refineries participated in the trading program in 1985 alone. Industry, initially uncertain about the value of trading, soon became an active user of the program. As an example of a well-utilized market-based environmental policy, lead trading was exceptional for its time and served as a point of reference for academics supporting tradable SO_2 permits in the 1990 legislation.

Generally, early market-based policies like netting and offsetting took a historical approach to the initial allocation of emissions rights by favoring prior users over new sources. The lead-trading program, however, was a surprising exception to this trend. All refineries received lead rights based on an identical rate per gallon of gasoline produced, regardless of their past production levels (Hahn 1989). Remarkably, the allocation did not vary according to the refinery's size or history of production and lead content. Total rights to use lead of course depended on the actual production of the refinery—the distribution equalized rates but not actual quantities of lead produced by each facility. But by providing equal rates to all firms, the lead program differed significantly from the allocation principles used by EPA in previous cases. The lead program relied on a more ahistorical, instrumental allocation that contrasts strongly with the rules initially proposed under the Title IV acid rain program. The allocation under the lead program thus anticipated the ideas of those who opposed a distribution in 1990 based on historical patterns of use.

Struggle Toward an Acid Rain Policy

The ecological problem of acid rain came to American policymaking prominence in the late 1970s.[4] Acid rain is actually the popular term for what scientists call acid deposition: the transport of atmospheric particles with a low pH via precipitation into lakes, forests, and human communities. It is caused by the emission of so-called precursor compounds, such as sulfur dioxide and nitrogen oxides, into the atmosphere through the burning of fossil fuels, especially coal. Precursors are not themselves acidic but change easily while airborne into chemical forms that include sulfuric and nitric acid. These powerful acids are then transported back to the earth's surface via rain and snow, frequently in locations hundreds of miles from the original pollution source. Acid deposition is suspected of causing ecological damage to lakes, fish stocks, and high-elevation forests. It is also linked to negative effects on human health and the accelerated deterioration of buildings, monuments, and historic structures (OTA 1985, 41–48).

Clean Air Act rules of the 1970s and 1980s were poorly designed for interstate pollution issues like acid rain. Indeed, by 1990 it was clear that the existing clean air law was exacerbating the acid rain problem in two respects. First, the act's emphasis on improving local air quality led many power plant operators to construct taller smokestacks to disperse pollutants across a larger area. This reliance on tall stacks increased the transport of SO_2 and other pollutants over long distances, intensifying the acid rain problem in areas downwind of large emitters, particularly the northeastern United States and eastern Canada.[5] In addition, the law's focus on regulating new sources of pollution was poorly suited to the acid rain problem. The law took this approach on the ill-founded assumption that older, dirtier plants were nearing the end of their economic lives and would soon be retired. By giving them a much weaker set of pollution control requirements, of course, the law encouraged utilities to extend the operating lives of these older plants. Utilities responded accordingly, and by 1990 most of the older, dirtier plants causing much of the acid rain problem were still in use (Kete 1992).

Dozens of bills circulating through Congress during the 1980s addressed the acid rain issue, inspiring intensive effort by lobbyists and lawmakers but meeting with little success.[6] Acid rain legislation raised contentious regional issues. Older, dirtier coal-burning power plants are concentrated in the Midwest and South, where prevailing winds carry their emissions into the Northeast and Canada. In addition, many of these plants burn midwestern coal, which is high in sulfur and therefore even more damaging in terms of acid deposition. Newer plants with much lower emissions (as required under the original Clean Air Act) are more common in western and some southern states. Low-sulfur coal deposits located in the West make that region eager to supply its coal to midwestern plants and reduce SO_2 emissions via fuel switching. These regional conflicts, combined with a lack of presidential interest and the powerful position of several members of Congress from midwestern and high-sulfur coal states, prevented enactment of acid rain legislation in the 1980s (Cohen 1995).

After years of rancor without progress, the landscape for acid rain legislation changed in the late 1980s. Senators Robert Byrd of West Virginia and George Mitchell of Maine, adamant in their opposition to and support for acid rain legislation, respectively, reached an agreement on the issue in 1988. Like its many predecessors, the Byrd-Mitchell proposal continued to take a command-and-control approach to the problem. It also contained an element of cost-sharing, in the form of a national utility tax to be paid by consumers in all 50 states. Representatives of states in the Midwest and South argued it was unfair for them to have to bear the full burden of cleaning up their older, dirtier plants and that other states should help shoulder the cost. The issue of cost-sharing to assist states with high-emitting utilities would remain a critical part of the 1990 debate as well. Although the Byrd-Mitchell proposal was not well received by environmental and other interest groups and therefore made little progress in Congress, it was an indicator that movement on the acid rain issue was likely to come sooner rather than later (Cohen 1995).[7]

After the 1988 election, acid rain legislation moved to the fast track. Mitchell successfully challenged Byrd for the Senate majority leadership post, eliminating one of the most effective institutional obstacles to acid rain legislation. In addition, George H. W. Bush replaced Ronald Reagan as president. Bush was publicly committed to being "the environmental president" and turned to clean air as an area in which he could immediately deliver a new proposal, perhaps in part because of the recent movement in Congress toward an agreement on the issue (Kete 1992). In addition, in 1987 the Environmental Defense Fund (EDF) had broken ranks with other environmental groups and begun touting an emissions trading proposal to address the acid rain problem. Bush became aware of the tradable permits idea for acid rain during his 1988 campaign, and it soon became a foundation of his bill to address the issue (Cohen 1995). With the new president's election, the paths of acid rain legislation and market-based policy finally intersected, eventually resulting in the Title IV portion of the CAAA, passed into law in 1990.

INITIAL PROPOSALS: EDF AND THE ADMINISTRATION

Almost as soon as he was elected, President Bush's administration went to work on a proposal to amend the Clean Air Act. Although a market-based mechanism to address SO_2 emissions was clearly going to be a part of the bill, the details remained uncertain. One critical question was how to allocate allowances; another was whether the bill would include a firm cap on total SO_2 emissions as part of the trading program. In this debate, the influence of Environmental Defense Fund staff as supporters for a trading program and a total emissions cap was vital; without their efforts, the chances are that a cap would not have been included in the Bush bill (Cohen 1995).

The SO_2 trading program started as an EDF proposal to the administration. An important innovation of the proposal was a change to total atmospheric emissions as the essential regulatory standard. Previous clean air law had focused on emissions rates only; total emissions by a plant or factory were unlimited as long as the rate for each facility met the appropriate legal requirements. In other words, plants could burn as many Btus of energy as they liked provided the emissions of SO_2 remained below a certain amount per million Btus. The problem with such a rate-based policy, of course, is that emissions reductions due to lower rates are eventually cancelled by growing fuel consumption and demand for energy. EDF proposed to shift the regulatory focus to the actual number of tons of SO_2 going into the atmosphere, thereby preventing today's emissions reductions from being lost to tomorrow's greater energy consumption. This shift to a "total atmospheric loading" approach also created the currency for the emissions trading program—a tradable allowance to emit one ton of SO_2.

Given EDF's focus on economics and environmental protection rather than distributive equity, it is not surprising that its initial suggestion for allocating allowances resembled a Humean approach. A strong, property-like right to emit was central to the plan. The rights created in the EDF proposal were called emissions reduction credits, rather than allowances, as they came to be known, but they were assuredly a secure form of licensed property. In fact, the proposal omitted the issue of future reduction or adjustment of these emissions rights without government compensation—a qualification of the allowance that became crucial in the legislative debate.[8]

The distribution of rights under the EDF plan was also Humean (see Table 3.1). Plants that currently met BACT standards and emitted at a rate lower than 0.5 pounds per million Btus were exempt from the program altogether and merely had to continue meeting existing legal standards.[9] All other existing facilities were given a series of decreasing allowance allocations over a 10-year period. The diminishing allocations were based on a fixed percentage of a plant's actual SO_2 emissions in 1985. By year 10 of the program, all plants not meeting BACT standards were to be allocated allowances equal to 40% of their 1985 baseline emissions, regardless of how low or high those 1985 emissions were.[10] New plants received no allocation and were forced to buy allowances from existing emitters at market prices. The initial EDF proposal thus embraced a proportional reduction from the existing distribution of SO_2 emissions, with no explicit discussion of the equity implications of such a choice. In other words, it is the first and last example in the acid rain debate of a Humean grandfathering of allowances based on the status quo under the typology described in Chapter 2.

After months of work, the president introduced his version of the bill at a press conference on June 12, 1989. The Bush plan moved away from the Humean qualities of EDF's proposal, offering tougher treatment of highly polluting power plants and weakening the new property right. The president's bill modified EDF's emissions credits by renaming them allowances

Table 3.1 Allocation Principles: EDF Proposal

Type of plant	*Fuel standard*	*Emissions rate*
High emitters	Status quo followed by pro rata reductions	Status quo followed by pro rata reductions
Low emitters	None	Status quo
New entrants	No allocation	No allocation

and explicitly permitting their future modification or reduction by the government without compensation.[11] The president also reduced EDF's allocation to high-emitting plants and rejected cost-sharing alternatives for these utilities in the form of any new electricity tax, as had been proposed in the 1988 Byrd-Mitchell proposal. Bush justified this position in terms of equity, arguing that a national tax violated the polluter-pays principle and would force electricity customers in clean areas to pay for sulfur dioxide reductions a second time.[12] His determined rejection of this form of cost-sharing was the opening salvo in the fierce exchange of equity arguments that dominated the next 18 months of debate.

Maintaining EDF's basic cap-and-trade model, the Bush proposal significantly revised the allocation of allowances (see Table 3.2). The president's team created a two-phase approach to controlling emissions. The first phase, starting in 1996, applied only to so-called big dirty plants—those with more than 75 MW of power output capacity and emissions rates greater than 2.5 pounds per million Btus.[13] As in the EDF proposal, plants in this group were allocated allowances based on their actual Btus of energy consumed between 1985 and 1987.[14] Unlike the EDF proposal, however, the Bush plan limited the allocations to the "big dirties" based on a fixed emissions rate of 2.5 pounds per million Btus rather than the actual 1985 emissions rates.[15] Given that these actual rates in some cases were 5.0 pounds per million Btu or higher, the Bush adjustment was an emissions reduction of 50% or more for many older, high-emitting plants.[16]

Thus, the Bush plan moved away from a strictly Humean allocation based on possession to a perspective including intrinsic and instrumental ideas. Within the equation for calculating allowances, one factor (fuel consumed) was based on a historical figure, but the other (emissions rate) was constant across the board. This difference seems arbitrary—why would intrinsic arguments be appropriate for one element of the formula and not the other?—until one considers the nature of each element with respect to property norms. Chapter 2 has already noted that intrinsic norms are strongest at their Lockean roots, when the prior use in question is widely viewed as beneficial and tangible. By itself, the consumption of fuel to generate electricity is close to this Lockean ideal: it represents work by utilities benefiting the larger community by providing a vital commodity. The emissions rate, however, is obviously much less Lockean. Here, the prior use is not fuel consumption but air pollution. Putting toxic compounds into the atmosphere is itself neither beneficial nor especially tangible. As a result, one would expect intrinsic norms to be much weaker with respect to this part of the allocation equation. The Bush proposal follows this expected pattern nicely, and future modifications to the allocation process would also focus on emissions rates rather than fuel consumption.

The second phase of the Bush proposal, scheduled to begin in 2001, also took an intrinsic allocation approach with instrumental qualifications. It addressed all power plants with a generating capacity greater than 75 MW and an emissions rate greater than 1.2 pounds per million Btus. As in the Phase I plan, these plants were assigned allowances based on their historical average energy input from 1985–1987 multiplied by a constant emissions rate of 1.2 pounds per million Btus.[17] Other than the lower rate, the method was the same as in Phase I for the big dirties. Smaller plants (below 75 MW in capacity) and cleaner plants (emitting at a rate lower than 1.2 pounds per million Btus) were exempted from the allowance system altogether and regulated under a structure similar to the old rate-based approach. They received no allowances at all and instead were limited to their actual 1985 rate of emissions, with no limit on the amount of fuel consumed.[18] In other words, unlike their big and dirty peers, small plants and clean plants were held to a historical (rather than equal) rate of emissions. In this respect, the formula for clean plants was the inverse of the formula for big dirties: the *rate* was

Table 3.2 Allocation Principles: Bush Proposal

Type of plant	*Fuel standard*	*Emissions rate*
High emitters	Historical	Equal
Low emitters	No limit	Historical
New entrants	No allocation	No allocation

based on historical use whereas the *capacity factor* was not. In addition to these basic rules, the administration proposal included a handful of exceptions for sources using special "clean coal" technologies to reduce emissions. New power plants still received no allocation under the bill; they were required to purchase their allowances from existing sources.

All participants in the proposed allowance program had specific obligations regarding monitoring of emissions and penalties for noncompliance. All sources, for example, had to install continuous emissions monitoring systems (CEMS) to measure their emissions at all times.[19] Sources emitting SO_2 without adequate allowances were to be fined $2,000 per ton of excess emissions and forced to give up allowances on a one-to-one basis in the coming year for every ton they exceeded their previous year's allocations.[20] These monitoring and enforcement rules remained unchanged in the bill until its final passage into law in 1990.

With the release of the Bush proposal, the acid rain policy debate shifted emphatically to questions of equity. The Bush team took policy ideas from EDF and other market advocates based on possession, security, and environmental protection and gave them a different basis. The administration rejected a grandfathering of allowances based on the status quo distribution of emissions, and the idea never returned to the bill. Instead, the president's proposal replaced the EDF allocation plan with distributive principles favoring prior use with some important instrumental modifications. That intrinsic tilt is consistent with the Hegelian pattern presented in Chapter 2, in which property rules start with the intrinsic perspective and then move in an instrumental direction over time. Although the Bush bill was far from an unqualified ratification of prior use, it was the strongest collection of intrinsic allocation rules to appear in the process, as Table 3.6 at the end of this chapter makes clear. Once the debate over acid rain entered the political realm with the Bush proposal, it was dominated by concern over these equity issues to the very end of the process. Congress would focus primarily on making instrumental adjustments to the Bush plan, as the next sections will describe.

CONGRESSIONAL HEARINGS: FRAMING THE DEBATE

Committees in both the House and the Senate held hearings on the Bush bill in the fall and winter of 1989–1990. In the Senate, both the Environmental

Protection Subcommittee of the Committee on Environment and Public Works and the full Energy and Natural Resources Committee participated. In the House, Indiana Representative Philip Sharp's subcommittee of the Committee on Energy and Commerce conducted extensive hearings on the bill.[21] All three sets of hearings devoted considerable attention to the bill's acid rain section.

Demands for fairness and equity permeated the proceedings. On the House side, Chairman Sharp raised the equity issue in the opening moments of the first session, stating that the committee must "... deal with the major inequity in this proposal—that is, allocating the burden of acid rain clean up."[22] Senator Max Baucus of Montana, chair of the Environmental Protection Subcommittee, also invoked equity in his opening statement by citing the need to redress "past wrongs that are deeply troubling to the West" under prior versions of the Clean Air Act.[23] On the administration side, EPA Administrator William Reilly described the bill as "fair" in his presentation to both House and Senate members.[24] Countless other speakers appealed to norms of fairness and equity in their testimony, as well as in question-and-answer exchanges with committee members.

As might be expected, those fairness arguments centered on allocation issues. The arguments about allowance allocation can be placed into four general categories:

- arguments about the type of right being created;
- allocation arguments in favor of high-emitting plants;
- allocation arguments in favor of low-emitting plants; and
- allocation arguments regarding access to allowances for new sources.

This section will describe the conflicts in each category as they appeared during congressional hearings, and the ways in which intrinsic and instrumental arguments were made by advocates and members of Congress regarding each of them.

The Type of Right Being Created

In contrast to the brief consideration given the issue in the Bush and EDF proposals, the question of whether an allowance was a form of private property was a source of great consternation in Congress. Many members of

Congress expressed confusion regarding the property status of allowances, especially with respect to future takings claims under the Fifth Amendment. Baucus, for example, asked proponents of a trading system about the takings question. In response, Senator Tim Wirth of Colorado assured his colleague that Congress could reduce or eliminate these allowances at any time without paying compensation under the takings clause.[25] Other testimony, especially from environmentalists at groups other than EDF, strongly reinforced Wirth's contention that these allowances were not a form of property.[26] David Hawkins of the Natural Resources Defense Council, for example, decried what he called "loose talk" that allowances are pollution rights. "There are no 'rights' to pollute," he argued, "nor should there be."[27]

Despite those assurances and the explicit language regarding this issue in the Bush bill, the view of allowances as a form of private ownership remained resilient. Members and advocates alike frequently seemed to subscribe to the idea that "if it walks like a duck, and talks like a duck, it's a duck." Senator Baucus asked his colleagues, for example, if allowances are to be bought and sold on the market, "do they then become property rights, by definition?"[28] Representatives of publicly owned power companies and others reiterated the view that the bill "conceive[d]" of allowances as private property rights, or at least as some form of private right, rather than as privileges.[29]

Others went a different route, agreeing with environmentalists that allowances were not property rights but arguing that they should be. As one utility executive stated, "[the bill] makes it clear that allowances are mere privileges, subject to cancellation at the whim of EPA. This falls short of the necessary assurances needed to support a commodity in an open market."[30] Some members of Congress were influenced by such arguments, despite their dismissal by administration staff and other supporters of the bill. Representative Bill Richardson of New Mexico, for example, proclaimed the property status of the allowances "rather unclear," while Representative Sharp worried out loud that the Midwest would have to sell allowances quickly "before the Congress has a chance to change the rules five years from now."[31]

The confusion over the property status of allowances was a direct result of their design. The bill's proponents wanted allowances to be secure enough to create a robust trading market, but not so secure as to require compensation to allowance holders in the event of future government regulation. Because of their ambiguous design, however, the property status of allowances has been an ongoing topic of debate to the present day. The uncertainty over the use of the term *property* in this case clarifies the usefulness of the licensed property idea. Instead of arguing at length over whether allowances are property rights or not, it makes more sense to define allowances at the outset as a form of licensed property encompassing many of the typical powers of ownership but subject to future government modification.

Arguments in Favor of High Emitters

Representatives of older, dirtier power generators in the Midwest and South fought for an allocation based on an intrinsic entitlement to historical emissions levels. Their initial arguments emphasized cost-sharing as a way of equalizing the burden caused by disproportionate reductions that might be required of them. This argument rejected a polluter-pays idea in favor of a quasi-Lockean entitlement to current emissions. Although they were based on present emissions levels, the arguments by high emitters were clearly not Humean. They lacked the focus on strong, secure property rights prominent in the EDF proposal, for example, as well as any reference to practical considerations or the "endowment effect" so central to Hume's account. Instead, the perspective was one of intrinsic entitlements for individual plants based on prior, historical use. Interestingly, these same high-emitting interests turned to instrumental arguments on other issues, such as inclusion of industrial sources in the program and modification of the historical fuel consumption baseline for some plants. These instrumental arguments were difficult to square with the basic intrinsic approach of the big dirties, however, and consequently failed to exercise much influence over the final bill.

Arguing that acid rain was a national problem similar to the then-recent savings-and-loan crisis, representatives of regions with older plants sought a national financial contribution to help them clean up.[32] United Mine Workers leader Richard Trumka, for example, declared flatly that the need for cost-sharing was based on "simple mathematics and the obvious need for equity in the formulation of environmental policy."[33] The argument was that the high-emitting states were being given an unfair burden: nine midwestern and southern states producing 50% of the current national SO_2

emissions would have to make 67% (or more, depending on the estimates) of the proposed reductions under the Bush bill. Sharp spoke for many others when he declared that any such disproportionate reduction beyond the 50% figure for these nine states was unfair. "[T]he fairest approach," he continued, "would be for each state to make reductions in direct proportion to its [current] emissions."[34]

Initially, supporters of cost-sharing sought a national electricity tax (similar to the Byrd-Mitchell approach in 1988) to pay for any disproportionate reductions. This idea was strongly supported by midwestern legislators, almost always on equity grounds.[35] The national tax was fiercely opposed by the Bush administration, however, as well as by numerous other constituents and members of Congress.[36] In opposing the tax, the administration also disputed the argument by high-emitters regarding their "excessive" share of emissions reductions. Acknowledging that certain states faced a larger share of emissions cuts, EPA staff and others argued that the costs of those reductions were lower per ton and therefore represented a proportionate share of the total expense of the SO_2 cleanup effort.[37]

In the face of such strong opposition to a national tax, an alternative version of cost-sharing based on the initial allocation of allowances gained importance. Although using allowances as the currency for compensation to dirty states was initially opposed by Representative Sharp and others, they soon became the coin of the realm. The issue of cost-sharing thus moved during the hearing stage from an issue of taxation to an issue of allowance distribution. It would remain a central part of the allowance allocation debate until the final conference committee meetings on the bill.

As cost-sharing began to be discussed in terms of allowances, the implied norms of entitlement based on conflicting theories of property became critical. The relevant rule in the administration proposal gave dirty plants allowances based on a rate of 1.2 pounds per million Btus (as of Phase II), while cleaner plants were held to their (lower) existing rates. High-emitting states did not view this "rate gap" as real cost-sharing, since it still required them to make larger reductions than other regions. The administration and many members of Congress saw the issue differently, however, calling the rate gap an important form of cost-sharing already in the bill.[38] Even Representative Jim Cooper of Tennessee, a state with historically high emissions, questioned the need for more cost-sharing, given the rate gap and other subsidies for high-sulfur coal already in the law:

> I realize that these folks who own the dirtiest plants want help in [cleaning them up], but when we are already planning on giving them two or three types of help, do we need a fourth type of help as well? To me that stretches the bounds of fairness.[39]

Conflict over the rate gap can be easily explained in terms of disagreement over norms of property theory. The clean states' perspective presumed that a distribution of allowances based on an equal rate of 1.2 pounds per million Btus for all plants was the equitable ideal. In calling the actual allocation under the Bush plan a form of cost-sharing, they implicitly rejected the prior use norms of dirty plants in favor of a more instrumental approach. Only on this assumption can deviations giving the Midwest and South more allowances than other regions through a higher emissions rate be considered a subsidy rather than an entitlement. Dirty states' representatives and others argued that holding them to a rate of 1.2 pounds per million Btus was no subsidy at all, since they were still being asked to reduce more than their current contribution to the problem. Their focus on proportional reductions from historical emissions levels implied an intrinsic norm of entitlement based on prior use (see Table 3.3). The valid measuring stick for fairness on this account was not a distribution of emissions rights based on a single rate for the entire nation.[40] Instead, a fair allocation in this view should be based on an equal *burden* of emissions reductions.

A surprising number of nonmidwestern interests, including many from the West and other clean regions, were at least partly sympathetic toward an entitlement to historical emissions levels, as long as it did not involve a national tax.[41] EPA Administrator Reilly, for example, argued that forcing new power plants to buy allowances from existing high-emitting utilities was fair because otherwise, these utilities would have to cut emissions to an even lower level. "[W]e've struck a reasonable balance," concluded Reilly,

Table 3.3 Allocation Principles: High Emitters

Type of plant	*Fuel standard*	*Emissions rate*
High emitters	Historical with pro rata reductions	Historical with pro rata reductions
Low emitters	Historical with pro rata reductions	Historical with pro rata reductions
New entrants	No allocation	No allocation

"and been quite fair in allocating the costs associated with this problem."[42] In defending the rate gap in this manner, Reilly and others implicitly acknowledged the equity of cost-sharing via an allocation favoring high-emitters.

A related fairness issue was the choice of a particular historical baseline for measuring fuel consumption. Dirty plants were granted allowances based on their actual fuel consumption averaged over three years, from 1985 through 1987. Numerous advocates and members of Congress decried the choice of year as arbitrary, unfair, and an example of regulation "by the roll of the dice."[43] (Such baseline issues in general soon became known as the "Class of '85" issue). High emitters took an instrumental approach in these arguments, as witness after witness explained why actual 1985 levels of fuel use were abnormally low and therefore unfair.[44] In seeking adjustments to 1985 figures, however, high emitters rejected the intrinsic allocation principles based on actual prior use that formed the very core of their cost-sharing position. If entitlements are to be set by norms of prior use, in other words, it seems unclear why those same norms should then be rejected just because some levels of prior use are lower than others. The high emitter's reliance on such instrumental arguments was an example of an inconsistent use of conflicting property perspectives—a position that proved hard to sustain in the ensuing debate.

EPA sympathized to some degree with the baseline complaints but defended the use of 1985–1987 data as the most current and accurate numbers available, upon which allocations should therefore be based.[45] This insistence on using the most accurate historical data reflects the intrinsic allocation norm that underlay many parts of the Bush bill. EPA staff also resisted wholesale flexibility in setting the baseline for fear of raising everyone's rates and allowances to a point that the environmental goals of the program would no longer be met.[46] Instead, they expressed willingness to make limited case-by-case adjustments to the baseline figures on fairness grounds, without specifying in the hearings which exceptions would be most persuasive to them. In this manner, they accepted the need to qualify but not reject their intrinsic allocation principle with some instrumental adjustments, in true Hegelian fashion.

Finally, high emitters sought one other important change to lessen their regulatory burden: including private industry in the provisions of the acid rain title. In 1990, 30% of all SO_2 emissions in the United States came from the private industrial sector. Yet for both political and practical reasons, the acid rain law placed restrictions only on utility-owned power plants. Utility facilities were far less numerous than industrial sources yet generated a majority of SO_2 emissions nationally. They were also a sector of the economy that was already highly regulated, making them an appealing choice for further control.

Many pages of the legislative record are devoted to statements attacking or defending the choice to exclude industry, frequently on equity terms. Sharp pointedly noted that all sulfur dioxide emissions have the same environmental impact regardless of their source.[47] This argument was supported by Cooper, who advocated the inclusion of industrial sources in the bill to reduce the compliance burden for dirty states.[48] In particular, representatives of the utility sector and their interests in the Midwest and South used polluter-pays rhetoric to make their case. Cooper justified the inclusion of industrial sources on just this basis, arguing that "those who cause the problem should bear the burden of solving it."[49] In other words, if the intent of the bill was to make polluters pay for the cleanup, as the Bush administration said it was, why were 30% of the polluters getting off free of charge?

The administration and large industrial sources resisted that argument with some appeals to fairness of their own. Industry stated that it would fund emissions controls through higher electricity rates and should not have to pay twice for this environmental effort.[50] EPA agreed with this argument, contending that the more equitable option was to make reductions "across the board on utilities and have the industrial sources pay their share of the utility bill."[51] The EPA argument relied in addition on the fact that industrial sources had significant new regulatory burdens under the air toxics portion of the law, and therefore were appropriately left out of the Title IV program.

By promoting the polluter-pays principle for industry, midwestern utilities and their brethren were again making inconsistent equitable arguments. The polluter-pays idea is fundamentally instrumental with respect to emissions rights: those who pollute the most have to clean up the most to obtain a nationally consistent standard. By using polluter-pays rhetoric regarding industrial sources, midwestern interests undermined the very principles that supported their larger share of allowances on intrinsic, prior use grounds. Having presented such a strong prior use argument in favor of sharing the burden of emissions reductions, it is hard to see how high-emitting plants could coherently justify an instrumental approach to determining fuel capacity figures or the inclusion of industry. For this reason, given the important influence of such norms in the process, it is not surprising that these instrumental arguments by high emitters largely failed to influence the details of the final law.

Arguments in Favor of Low Emitters

The Bush bill treated clean power plants very differently from their high-emitting peers. As noted above, any plant currently emitting below the rate of 1.2 pounds per million Btus was limited to its actual 1985 emissions rate indefinitely. Existing plants were permitted to increase capacity, but only so long as they maintained their 1985 rates. During the hearings, clean states protested these limits on existing plants, as well as the rate gap between clean plants and dirty ones, largely on instrumental principles. Ironically, environmentalists in turn rejected the arguments of the clean states for higher emissions rates on instrumentalist grounds as well.

Although some representatives of regions with low-emitting plants sympathized with the rate gap, others did not. Consider this acerbic exchange, for example, between Representative Howard Neilson of Utah and Paul Centolella, appearing before the committee representing the high-emitting state of Ohio:

> Nielson: ... if [Ohio power plants] do go from [an emissions rate of] 1.2 down to 0.9, you do get credit at least for the difference between 1.2 and 0.9, but if we go from 1.2, which is our dirtiest [plant], to 0.9, we get no credit at all. Is that equitable?
>
> Centolella: Even under an emission—
>
> Nielson: [interrupting] Is that equitable?
>
> Centolella: No, it is not equitable.
>
> Nielson: Thank you.[52]

Time and again, clean states argued that they had already controlled their emissions as required under previous versions of the law and deserved allocation rules that recognized this effort. One representative from a clean state dismissed the historical argument for allocation in a twist on Lockean terminology, stating that allowances should not be allocated based on a "right of prior abuse."[53] Most of these complaints instead sought an allocation using the same 1.2 pounds per million Btus rate for all plants, or at least some movement in that direction (see Table 3.4). The instrumentalist property perspective embodied by these arguments is clear.

Beyond their vociferous attacks on cost-sharing, some clean states also argued for allowances based on growth. Representative Michael Bilirakis, from the rapidly growing state of Florida, was very hard on EPA Administrator Reilly on this point. "[F]or the life of me," he concluded, "I don't understand what is fair about this, and why should the customers of a relatively clean utility have to buy allowances from dirtier utilities to accommodate any growth?"[54] The administration's response—that clean plants could increase their capacity as long as they remained at the same emissions rate—was unconvincing to politicians and clean power plant operators alike.[55] Eventually, these fairness concerns led to serious changes in the law's treatment of clean plants.

In an example of politics creating strange bedfellows, environmentalists found themselves agreeing with the "big dirties" in opposing more allowances for clean plants. Fearing the deterioration of pristine air quality in many rural western areas, they were reluctant to acknowledge any entitlement to higher emissions rates for the power plants within those states.[56] It is worth noting that this argument opposing the clean states was also an instrumental position. Rather than seeking a more equal distribution of allowances, however, environmentalists wanted an allocation that met other social goals, such as preserving the clean air in many national parks and other scenic regions of the West. In the face of this national priority, their position implied that inequalities regarding emissions rates were of secondary importance.

Table 3.4 Allocation Principles: Low Emitters

Type of plant	*Fuel standard*	*Emissions rate*
High emitters	Historical	Equal
Low emitters	Historical with additions for growth	Equal
New entrants	No allocation	No allocation

Like high emitters, low emitters also had a complaint about the bill's 1985 baseline. In this case, however, the issue concerned emissions rates rather than fuel consumption. The bill limited clean plants indefinitely to their actual 1985 emissions rate. As in the case of fuel consumption for dirty plants, some clean operators argued that their actual emissions rates were atypically low that year and therefore unfair as a permanent emissions standard.[57] An additional complaint arose over the use of actual rates rather than legally permitted ones (which were typically higher) as the new legal limits. This rule seemed to penalize clean plants that overcomplied with the law in 1985 by giving them fewer emissions. A solution to this particular Class of '85 problem for clean plants, at least, was to entitle them to their legal, rather than actual, 1985 emissions rate. The idea was proposed by several witnesses to the hearings;[58] one power company executive put it most clearly:

> It doesn't seem to me that it makes sense to make a performance of 0.9 [emissions rate] illegal when the permit based on well-defined criteria allowed 1.0. Such plants should not be forced to find additional reductions or reduce their operating levels when they are the cleanest plants in the country.[59]

The debate over a permitted versus historical standard for emissions rates is another good example of a clash of intrinsic and instrumental ideas. By proposing actual emissions rates for clean plants in 1985 as the fair standard, EPA and the administration effectively argued once more on intrinsic grounds that historical behavior provided the equitably relevant data. Clean plants arguing for emissions at their permitted rate relied on the instrumental view that regulation, rather than actual use, created an effective entitlement to future emissions at the same limit. Thus, clean plants again protested against a prior use allocation on instrumental terms, much as they contested the rate gap favoring historically high-emitting plants in the Midwest. When made by clean plants, however, these instrumental arguments against historical emissions rates were fully consistent with their arguments on other points. They also addressed the part of the allocation equation (emissions rates rather than fuel consumption) most amenable to instrumental modification in terms of distance from the Lockean ideal of beneficial and tangible use. Not surprisingly, those arguments were therefore more successful than the instrumental positions taken by their high-emitting peers.

Allowances for New Sources: Access and Allocation

Despite the many instrumental arguments being offered in the hearings, the issue of new emissions sources offered a good example of the continued power of the intrinsic property perspective. The Bush proposal included a 100% offset requirement for all new sources; that is, any plant built after the law was passed would have to buy allowances for *all* its SO_2 emissions from existing sources. This requirement was included to maintain the program's firm cap on total SO_2 emissions, an aspect of the bill dearest to many of its supporters. The cap itself was a controversial part of the law, and some speakers in the hearings denounced it as ill conceived and unnecessary. Beyond this point, however, the issue of how new power plants were to obtain allowances given the cap's existence was a central, and more thought-provoking, issue.

Fear of "allowance hoarding" by high emitters dominated this discussion. States expecting high rates of growth, for example, worried that dirtier midwestern utilities might refuse to sell their allowances.[60] The fear of hoarding was echoed by representatives of the new independent power producers (IPPs), who intended to build energy-generating capacity to compete with existing utility sources. The issue was repeatedly presented at the hearings as one of ability to purchase allowances from high-emitting utilities.[61] Those defending the Bush proposal and the proposed market in SO_2 allowances spent a great deal of time reassuring members of Congress that the market would function properly and that new sources would be able to buy allowances from older plants that had overcontrolled their emissions.[62] IPPs and representatives from growth states were less confident of the market's liquidity; their concerns eventually led to inclusion in the bill of both an auction and a direct sale of allowances to ensure access for new sources.

What is remarkable about this dialogue is the nearly universal understanding of the issue in terms of *access* to allowances rather than *entitlement* to them. Some testimony from growth states and clean states did seek a limited entitlement to emissions rights for very clean new plants by trying to exempt them from the offset requirement.[63] Such arguments focused on eliminating or weakening the cap, however, rather than on creating a redistribution of allowances for these new plants. IPP representatives refrained in their committee appearances from offering any argument for a direct allocation of allowances to their new plants.[64] One IPP representative focused so strongly on access rather than entitlement that he described the failure to compel trading (rather than reallocating allowances directly to IPPs) as the "basic flaw" of the proposal.[65]

A reallocation of rights to new sources was clearly feasible in terms of the mechanics of the program; allowances that were withheld from existing sources for auction could just as easily have been redistributed to new sources at no cost. It may not be surprising that Congress failed to include this kind of redistribution of allowances in the bill, since IPPs were not a particularly strong lobbying force in 1990. It is surprising, however, that IPP representatives and others did not even mention such a redistribution of rights during the vast majority of their testimony. This acceptance by IPPs and others that new sources would have to buy their allowances is a strong indicator of the power of the intrinsic, prior use norm. Even those who would have substantially benefited from an instrumental perspective giving emissions rights to new sources failed to challenge this historically based aspect of the allocation scheme. There is no better evidence of the power of the intrinsic position than this omission.

Summary of Hearings Arguments

The hearings on the CAAA paid close attention to the acid rain portion of the bill. Fairness issues were of primary concern, as was the question of whether allowances were a form of private property. In terms of allocating allowances, the intrinsic-instrumental axis of property ideas was central to the debate. The original bill set the stage by taking and defending an approach tilted toward intrinsic rights on most allocation issues. This intrinsic starting point for the allocation debate matches the first step of the Hegelian pattern for creating property rights, as described in Chapter 2.

Midwestern and southern utilities sought an even stronger intrinsic allocation based on prior use, in the form of a national tax to help clean up high-emitting plants. As the cost-sharing debate shifted to allowances, however, there was growing acceptance of the alternative view that the bill already included cost-sharing by giving dirty plants a higher emissions rate than clean ones. Clean states argued against the bill's initial allocation rules, including the rate gap, on instrumental grounds. The next section will show how these arguments eventually succeeded in reshaping the bill without destroying its original intrinsic basis, thereby following the Hegelian pattern. New sources were surprisingly accepting of the intrinsic principles of the Bush allocation that excluded them from any allowances except via purchase. High emitters continued to urge the inclusion of industry and changing the 1985 baseline on instrumental grounds. These arguments fared more poorly, however, in part because they were so strikingly contrary to the fundamental principle underlying the high emitters' position: that allocations should be based on intrinsic rights to historical levels of emissions.

CONGRESSIONAL DEBATE AND THE FINAL STATUTE

By January 1990 most of the hearings on the bill were concluded. Senator Mitchell's position as majority leader, along with President Bush's continued support, made the CAAA the top legislative priority for the coming year. The Senate Environment Committee reported a version of the bill late in 1989, and it became the first topic on the agenda of the full Senate after the winter recess in January 1990. After a lengthy deliberation, the Senate passed a version of the bill on April 3, and the House approved its own version in May. The conference committee to resolve differences between the two bills did not start work until July, however, and was soon interrupted by a two-month congressional recess. Despite intense pressure from both houses to finish the bill well before the November elections, the conference committee did not complete its negotiations until the early hours of October 22 (Cohen 1995). After both houses approved the conference bill, President Bush signed the act into law on November 15, 1990, nearly 18 months after he first introduced the bill.

The process during congressional debate and amendment of the bill was much less open than the hearings. Many discussions took place in closed-door meetings in both the House and the Senate, rather than on the floor of either chamber (Joskow and Schmalensee 1998). No public records exist of many of these private meetings, but interviews with individuals who were involved in the deliberations provide perspectives missing from the written record. This section will rely on both written records (including the final statute) as well as personal interviews in discussing the role of equity norms in this phase.

That equity norms remained central in shaping the final law is clear. Frequent references to the perceived fairness or unfairness of the bill continued as the bill proceeded through Congress.[66] This rhetoric is once again reflected in the law itself, the final version of which devotes more than 20 pages to equity-based distributional rules for allowances.[67] Interviews with persons closely involved with the bill confirmed that disputes over allowance allocation were central and that many arguments regarding those allocation issues were framed in terms of equity concerns.[68] Former EPA staffer Nancy Kete (1992, 80) also has noted the emphasis on equity:

> [A]lthough the emissions allowance trading system was designed [by bureaucrats] to encourage cost-effectiveness, during the Congressional debate itself, the overall environmental effectiveness of the programme *and the programme's fairness* were the members' pre-eminent concerns, not its economic efficiency or its potential cost-effectiveness [emphasis added].

Thus, when senators such as Pete Domenici of New Mexico trumpeted the fairness of the final bill coming out of the conference committee, they were not just paying lip service to the term.[69] Others certainly disagreed with Domenici's rosy assessment of the final product, but the senator's emphasis on the importance of equity in the process appears to be quite accurate.

The equity aspects of the 1990 legislative debate over the bill built directly on the allocation issues first raised in the hearings. Generally, members of Congress tried with mixed success to weaken the intrinsic allocation rules and policies in the administration bill with instrumental modifications. The result was a bill that eventually took a more Hegelian approach to allocation, instrumentally modifying some intrinsic principles but not others in crafting a "fractious holism" of the two ideas. Because of this continuity of issues, this section will follow the same organization as the previous one.

The Type of Right Being Created

Discussion over the type of right being created under the CAAA continued to be a significant part of the congressional dialogue. The Senate committee report on the bill contains extensive discussion of the property status of an allowance, as do the floor debates over the conference bill. Several interviewees also confirmed that the property status of allowances was an important point of consideration.[70] The result was a statute creating a clear example of a licensed property right with assurances, but no legal guarantee, of security against future reduction or elimination by government.

The Senate version of the bill revised and strengthened the original Bush language on this issue, adding the statement, "Such allowance does not constitute a property right."[71] The Senate committee report clarified that this clause was included in the bill to protect the government against takings claims under the Fifth Amendment in case allowances were later reduced or revoked.[72] For some members and advocates, including at least three Nobel prize–winning economists, the weakened property status of allowances threatened proper functioning of the proposed market.[73]

Because of those concerns, Congress tried to assure utilities and others that they could still rely on the security of the new emissions rights. The Senate committee report, for example, followed its rejection of the allowance's property status with a stated desire to create "economic commodities" in the emission of SO_2. "Accordingly," the report concluded, "allowance holders should expect that allowances will partake of durable economic value and that...relevant laws will apply to allowances and function to protect that value."[74] This philosophical straddle on the security of allowances was not unique to the committee report. Senator Baucus made a similar point, stating that allowances would be treated "in part" like commodities, but not to the degree that they would be protected by the Fifth Amendment.[75] Representative Michael Oxley of Ohio agreed, stressing that allowances "do have durable economic value":

> In short, while the Congress will not compensate holders of allowances for the loss of an individual allowance, the holders and users of allowances will have every reason to rely upon the continued existence and value of those allowances as they design and undertake their compliance efforts.[76]

The battle over the term *property right* to describe allowances thus continued from the hearings into the legislative deliberations of Congress. Despite the direct language of the final statute, members never fully clarified the degree to which an allowance could accurately be described as a form of private property. Allowances clearly are not protected by the Fifth Amendment but also clearly embrace many powers typically ascribed to private ownership. The language of the debate and the final law continue to support the vision of an allowance as a licensed property right, even as legislators struggled to come up with a term to describe the new rights they were creating.

Allowances for High Emitters

During congressional hearings, it became increasingly apparent that there would be no national electricity tax to fund cleanup costs at high-emitting power plants. With the resulting focus on allowances as the mechanism of cost-sharing, Congress amended the original bill in a few ways favorable to dirtier power plants. These successful forms of cost-sharing took an instrumental rather than intrinsic approach. Unlike other, unsuccessful instrumental arguments by high emitters, however, the successful ones managed a much better fit with the fundamental, intrinsic norms underlying the positions of this group and the bill in general.

The first step toward additional cost-sharing was to make more allowances available for redistribution under the total cap. The fixed political goal of a 10 million-ton reduction in annual SO_2 emissions made redistribution difficult—where were the new allowances for high-emitting utilities to come from? Congress solved the problem creatively by accelerating the compliance dates of the program. Requiring utilities to meet their new goals a year earlier than in the president's proposal created for the first time a substantial, one-time pool of allowances for redistribution.[77]

The biggest portion of this new "bonus pool" of allowances was assigned to high-emitting power plants that adopted scrubbing technology before the law's 1995 deadline. As long as they reduced their emissions by at least 90% through scrubbing, these plants would be eligible for a share of the additional allowances.[78] The result of a proposal by Senator Byrd's office, the pool created more compensation for high-emitting utilities and helped protect the high-sulfur coal mining industry in states such as West Virginia (Cohen 1995). Although the pool failed to satisfy fully the fairness demands of high-emitting states, it was clearly a move toward greater cost-sharing.

The bonus pool allocation was based on an odd mixture of intrinsic and instrumental principles. As a redistribution to previously dirty utilities, it moved away from the instrumental allocation standards sought by clean states. However, it was not simply an extension of intrinsic, historical allocation principles. Through its scrubbing requirement, the pool augmented allowance shares based on future actions by high emitters without contesting the basic historical allocation for this group. In this manner, it represented an instrumental, class-based rule that (unlike polluter-pays arguments) was consistent with the original allocation approach sought by the high emitters. The parallel to the Hegelian model again stands out.

In contrast to the success of the bonus pool, instrumental arguments by high emitters against the 1985 baseline for fuel consumption fared poorly. Unlike the bonus pool, the baseline complaints went against the historical allocation principles supported by high emitters. Rather than providing a wholesale adjustment in favor of plants that were operating well below capacity in 1985, Congress provided only minimal relief for a few specific units affected by tornadoes and other natural disasters.[79] Otherwise, only new plants that were just beginning energy production in the early 1980s were provided some adjustments from their actual energy use figures.[80]

Arguments for including industry in the acid rain program met a similar fate. During floor debate, midwestern politicians like Senator Alan Dixon of Illinois continued to press this point:

> I know the opponents of these amendments said that we should not be paying dirty states to clean up their problem. Well what about all of the dirty states that are let off the hook entirely, some of which actually will increase their emissions? Should they not help share the burden? Mr. President, I am just trying to point out that if we are going to call some states dirty because their sulfur emissions happen to come from utilities, and call other states clean because their emissions come from industrial sources, *we are going to have a fairness problem* [emphasis added].[81]

The senator went on to note that Texas, for example, would more than double its annual emissions of SO_2 under the bill despite being the third-largest emitter of sulfur pollutants in the nation at the time of the bill's passage.[82] Surely, he concluded, such discrimination was arbitrary and unfair.

The continued protests of the electricity sector and the Midwest notwithstanding, however, industry has no mandatory role in the Title IV program under the law.[83] The polluter-pays argument for including industrial sources remained in conflict with their basic position regarding greater allocations for utilities with the highest pollution levels. It is suggestive that in this case, as in the case of adjusting fuel baselines above, instrumental arguments lacking a certain level of equitable consistency with intrinsic principles failed to become part of the law. The failure of these arguments fits a pattern described by several interviewees, who noted that the most logically coherent equity arguments regarding allocation were often the most effective.[84] Given that midwestern arguments to include industrial sources were not consistent with the region's own arguments against the polluter-pays principle elsewhere, their eventual failure in the process is not surprising.

Allowances for Low Emitters

Congress made several important adjustments to the allocations for low-emitting plants in so-called clean states. In each case, the adjustment increased the allocation to certain utilities on equity grounds. In particular, the administration's attempt to limit low-emitting plants to their historical emissions rates was rejected by a Congress seeking other principles by which to allocate allowances. The instrumental vision of property was crucial to these changes.

The first change for low emitters was the decision to include them within the system of allowance allocation. Recall that the original administration bill kept clean plants out of the allowance system altogether, simply holding them to their 1985 rates in perpetuity. House members first recognized that these utilities could effectively increase their emissions by joining the allowance system. Representative Sharp offered a successful amendment to the bill in his committee to make this adjustment.[85]

Having become a part of the allowance system, clean plants then tried to grab a larger slice of the pie. The Sharp amendment gave clean utilities a 40% increase in allowances from their historical fuel consumption levels as a cushion for future growth. In the Senate, a similar adjustment was made with an increase of 20% (the final version of the law used the Senate's 20% figure[86]). Other rules in the final statute also adjusted allocations in favor of low emitters.[87] A "ratchet" in the law maintained the allowance cap despite these adjustments by reducing all allowances proportionally to the original total once the special allocations had been calculated.[88]

The adjustments favoring low emitters were a clear departure from an intrinsic standard of allocation in favor of instrumental rules. Nevertheless, as with the dirty states, adjustments that respected the intrinsic basis of the initial bill fared the best. In particular, the 20% growth adjustment represented a share-based rule creating greater equality between clean and dirty plants without erasing the previous advantage enjoyed by large prior users. As a pro rata adjustment of historical levels of fuel consumption, it was an instrumental rule that remained grounded in the original, intrinsic allocation approach of the bill.

Despite the changes favoring states with lower emissions and higher growth by modifying the prior use allocation principle, discontent among members of Congress from clean states remained significant as the bill moved toward the conference committee.[89] Some of their biggest outstanding concerns revolved around baseline issues and the use of actual emissions rates to calculate the allowances for clean plants. Once again, the instrumental arguments of clean sources fared better in Congress than those of dirty ones. After months of debate, the bill changed in the conference committee to provide allocations for low-emitting sources based on their *permitted* rather than *actual* rate of emissions during the baseline period.[90] The adoption of the permitted emissions rate for calculating the initial allocation is an innovative example of a share-based rule, distributing allowances based on legal entitlements rather than on actual levels of use.

The new baseline rule continued the pattern of instrumental arguments' having more influence on the emissions rate in the allocation formula than on the fuel consumption figure. Attempts to shift the historical baseline for fuel consumption were largely ineffective, except for those making a proportional, pro rata shift based on the actual 1985 figures. In the final version of the bill, allocations remained largely based on historical levels of fuel consumption (see Table 3.5). By contrast, the emissions rate never relied as much on historical figures even in the initial bill and moved even further away from a historical standard with the shift to permitted rather than actual rates for clean plants. The different nature of the prior use in burning fuel versus emitting pollutants helped determine the relative power of the intrinsic, Lockean allocation norm for each part of the formula.

Through adjustments like the 20% growth factor from the baseline and the use of the permitted emissions rate, supporters of plants with a history of low emissions used instrumental principles to increase their share of the distribution. Meanwhile, dirty plants were far less successful in adjusting their allocations based on instrumental arguments. The resulting distribution was still a far cry from an egalitarian one; historically dirty plants received many more allowances than clean ones even after the changes made by Congress. Nevertheless, the instrumental approach was a much stronger part of the bill after Congress finished work on it (see Table 3.5). Clean states made significant gains based on these changes.

Table 3.5 Allocation Principles: Congressional Outcome

Type of plant	*Fuel standard*	*Emissions rate*
High emitters	Historical	Equal
Low emitters	Historical with 20% pro rata increase	Legally permitted
New entrants	No allocation	No allocation

Allowances for New Sources: Access and Allocation

Throughout the legislative process, the dominant issue for new sources remained market access. Congress responded by creating two mechanisms for ensuring liquidity in the allowance market: a direct sale and an auction.[91] The direct sale guarantees a limited number of allowances to be withheld from existing sources for purchase by new power generators at a fixed price of $1,500 per ton. The auction also withholds allowances from existing sources for EPA to sell to the highest bidder on an annual basis. Each year 2% of the total allocation of allowances is withheld for these two market enhancements, with revenue from both sales being returned directly to the former owners of the allowances. The auction and direct sale provisions are an effort to ensure market access only; they have no aspirations to redistribute wealth.

During the congressional consideration of the bill, independent power producers remained quiet about asking for a share of the initial allocation. Market access was the issue; the option of an initial redistribution to IPPs was never on the table.[92] The bill did eventually bequeath a modest number of allowances to a handful of power plants that were actually under construction at the time of the bill's passage.[93] Other than this modest exception, the principle of allocation to prior resource users went unchallenged. Even during the more private discussions outside the hearing process, IPPs and members of Congress apparently continued to avoid any serious suggestion of allocations in violation of the prior use idea. Here was an area, like the question of including industry, in which instrumental ideas made no headway at all in changing an intrinsic approach of the original bill.

Summary of Legislative Arguments

The final version of the bill contained several important instrumental adjustments by Congress. The property status of an allowance became less secure, despite continuing congressional assurances that allowances were to be a reliable and fully tradable asset. Several instrumental arguments in favor of modifying intrinsic allocation principles succeeded, primarily though not exclusively to the benefit of clean states. Instrumental arguments made by high-emitting plants and states were less successful, in part because they more frequently lacked coherence with the intrinsic principles that underlay most of the original bill. In particular, instrumental norms were much more influential over one element of the allocation formula, the emissions rate, than they were over the other element, fuel consumption. This unequal influence supports the hypothesis that intrinsic norms will be more influential when the prior use in question is widely seen as beneficial and tangible.

In the end, Congress modified the Bush bill in an instrumental manner while leaving certain core intrinsic principles intact. That instrumental arguments were used to revise but not contradict or overturn the strong intrinsic principles of the bill accords with the dialectical process required by Hegelian property theory. The outcome was much more of a fractious holism than an allocation based on intrinsic rights alone or a grandfathered distribution based on the status quo.

IMPLEMENTATION ISSUES

The allocation of allowances under the acid rain program was primarily a legislative process. In fact, Congress provided a degree of clarity and specificity in the CAAA regarding the allocation of emissions rights that was uncommon for a modern environmental statute.[94] As a result of this detailed legislative process, the administrative and implementation issues for the statute have largely focused on areas other than allocation. Nevertheless, a few distributive equity issues have surfaced since the law's passage, including more controversy surrounding the exact property nature of an allowance. Other important implementation issues have involved the allocation of the bonus pool of allowances for early adopters of scrubbing technology, the auction and direct sale provisions of the law, and the issue of geographical concentrations of emissions, or "hot spots." This section will briefly cover each of these issues.

What Is an Allowance? The Debate Continues

Congress tried to be clear that allowances were not a vested property right subject to compensation under the Fifth Amendment. At the same time, it wanted allowances to have sufficient security to stimulate a robust trading program. The early indications are that the philosophical straddle on the issue has worked reasonably well: trading has been extensive under the program, with the volume of trades increasing nearly every year.[95] Yet the price of allowances has remained fairly low: less than $200 per ton versus predicted levels of $500 per ton or more (U.S. EPA 2000). The relatively low price of allowances may derive from more than one factor, including much lower costs of reducing SO_2 emissions than were originally predicted by industry. It may also reflect the relative insecurity of these licensed property rights. Prudent utility managers recognize that EPA or Congress may reduce the value of these allowances at any time. The greater the perceived threat of this type of reduction, the lower the market price of an allowance will be. At least one utility representative felt that the low market price of allowances was due in part to this issue of security. The relative security of allowances therefore remains an ongoing implementation issue for the program.

In addition, the valuation and taxation of allowances have raised the property question in new ways. Utility regulators have struggled with how to treat income from the sale of allowances. Initial rules typically treated allowances as though they had no value for utilities until they were sold, at which point the revenue was considered income and taxed as a capital gain (Burtraw and Swift 1996). Other utility regulators required that any income from the sale of allowances be refunded directly to ratepayers rather than shareholders, implying that the public "owns" the emissions rights (Bohi and Burtraw 1997). This process may be inhibiting the market, since utilities pay the tax penalties and lose the value of allowances only when they actually sell them.

The view that allowances have no intrinsic value and are pure profit when sold contradicts the entitlement arguments made by utilities before Congress. If regulators shared the utilities' view of allowances as an intrinsic form of licensed property, they would allow utilities to keep their income from the sale of allowances, rather than treating it as a windfall to be refunded to ratepayers. Utilities and other market advocates are seeking to change these accounting rules to stimulate trade and implicitly strengthen the property nature of these emissions rights.[96] In the controversy over accounting and taxation, the exact property nature of the allowances and the proper basis for allocating them remains an important issue.

Cost-Sharing: The Bonus Pool Dilemma

The bonus pool of allowances for units adopting scrubbing technology became the subject of its own allocation controversy during the implementation of the program. The law was vague on how to distribute the pool should there be more qualified applications than allowances.[97] One possibility was a strategy of first-come, first-served, giving a full set of allowances to each qualified applicant as requested until the pool was depleted. The other was to pro-rate the allowances among all qualified applicants as of a certain date. EPA thought that the law did not permit a pro rata distribution and in the end chose a third option—to allocate the bonus allowances on a lottery basis among qualified applicants. Utilities, however, made the EPA decision moot by agreeing to pool the allowances and redistribute them among themselves on a pro-rata basis, regardless of how EPA decided to allocate them.[98]

This egalitarian allocation of the bonus pool allowances by utilities is an interesting contrast to the prior use principles advanced by dirty plants burning high-sulfur coal. In this case, the notion of prior use or first-come, first-served was less compelling, perhaps because everyone involved was adopting the required technology within a few years of one another. Because of the synchrony, a shift to a much more instrumental and egalitarian distribution strategy among high-emitting sources was logical. Here is another example, then, of an instrumental rule modifying the prior use aspects of the allocation to dirty plants.

The Access Issue: Auction and Direct Sales

The problem of access to allowances has failed to materialize. No hoarding has occurred, and EPA cancelled the direct sale program for a lack of interest (Grant 1998). Clean states, so concerned about acquiring new allowances, have often been net sellers of SO_2 emissions during the first phase of the program.[99] Private brokers and other middlemen have entered the market and created greater liquidity; private trades between utilities and these brokers are widely reputed to be a substantial part of the market that goes untracked by EPA.[100] The auction has been more important than the stillborn direct sale program, but primarily as a price indicator rather than as a method of providing allowances for new entrants to the program (McLean 1997). Existing rather than new facilities have been the primary buyers in the EPA allowance auctions. Others, such as environmental groups, have participated in auctions but have bought only a tiny fraction of the total allowances being offered for sale.[101] Direct allocation of allowances to new sources continues to be a nonissue.

Indeed, the only way in which a redistribution of allowances to new sources seems imaginable is if Congress reduces the total number of SO_2 allowances available. Any further reduction in the total permitted loading of SO_2 in the atmosphere could have two types of distributional impact. More likely would be a proportional reduction in the value of allowances held by current emitters—a two-for-one allowance swap, for example. Less likely but certainly possible would be a reassessment of the baselines and other standards for allocating allowances as part of a reduction in the total cap. Some interviewees mentioned support for this type of reallocation of rights in the future, which would likely result in more allowances for sources that were not in operation when the 1990 law was passed.[102] Any such change would reopen the allocation issue in general and might have significant effects on the perceived security of the emissions allowance as a form of licensed property subject to saving and trading.

Hot Spots: Distribution of Burdens Rather Than Rights

An important equity issue beyond the initial distribution of allowances is the eventual distribution of pollution impacts under the law. Other emissions trading programs directed by EPA have led to serious concerns about the concentration of pollutants in certain communities. The RECLAIM program for trading emissions of air pollutants in Los Angeles, for example, has been accused of concentrating toxic pollution in minority and poor communities.[103] By permitting companies to trade freely, the program may force neighbors of facilities that are buying credits to suffer additional negative health and environmental impacts. Charges of environmental racism are a part of this criticism, and EPA has been sued over alleged hot spots in the Los Angeles program (Chinn 1999). Concern about the possible concentration of pollution is thus a leading equity-based criticism of emissions trading.

The acid rain program has been the subject of some litigation and controversy over the concentration of emissions allowances in the industrial Midwest.[104] A pattern of trading in which clean plants sell permits to high-emitting midwestern utilities could then intensify the effects of SO_2 emissions on northeastern areas that are downwind of these plants. New York State and environmental groups such as the Adirondack Council have been vocal about these concerns, but to date such concentrations of SO_2 emissions in the Midwest have generally failed to materialize (*New York Times* 2002; Burtraw and Swift 1996). Thus, trading in the acid rain program has not yet created the same intense controversy over pollution hot spots that marks the Los Angeles program. However, the possibility that future trading may lead to concentration of acid deposition in fragile areas like the Adirondacks remains a potential equity-based implementation issue.

CONCLUSION

The CAAA acid rain program created a new form of private ownership that is well described by the idea of licensed property, introduced in Chapter 1. The property status of allowances to emit sulfur dioxide was a major point of discussion during the passage of the legislation and remains a significant issue during the program's operation. Allowances are clearly not a vested right of property requiring government compensation if revoked, and yet they are also clearly intended to embody most of the powers of private ownership. As exclusive, fully tradable, and relatively secure rights to emit SO_2, allowances are a perfect example of licensed property rights and should be studied and discussed as such.

Parties on both sides of the debate may be reluctant to accept this particular definition of allowances, since it fails to match precisely their own positions. Yet in the end, protests that allowances are not a right to pollute are off-base: allowances are explicitly designed to be like property rights to promote market transactions, and distaste in some corners for the term *prop-*

erty does not alter this fact. Similarly, allowances are clearly not fully vested property rights under the law, either, despite the wishes of some utility executives and others that they were. The licensed property idea might help defuse such controversies over the property status of allowances at the outset by introducing a new term more descriptive of the instrument. It also helps analytically by permitting the use of property theory to understand the allocation process, bringing the Title IV program into the family of similar government policies using licensed property, as reviewed in Chapter 1.

Equity norms played a central part in the law's design and enactment. A review of the legislative history of the acid rain portion of the bill, from proposal to adoption and implementation, reveals that equity concerns were central to the process, particularly to the issue of how to distribute the initial allocation of allowances among affected utilities. Numerous participants in the 1990 process confirmed that allocation questions dominated legislators' attention on the matter, and that the most successful arguments for revising a specific allocation rule were rooted in equity terms. Despite the protestation by Senator McClure of Idaho that "... there is no complete or probably never will be any complete equity in this bill," the majority of congressional attention went toward trying to achieve that difficult goal.[105] A large section of the resulting statute is composed of complex rules that address these equity-based distributional concerns.

Nevertheless, much of the literature on the acid rain program refers to the initial distribution of rights created by these rules simply as grandfathering—a misleading description of the outcome. This chapter has demonstrated in great detail the wealth and complexity of the equity-based arguments that went into forming the initial allocation. Grandfathering, or a Humean distribution based purely on status quo emissions, was never a serious part of the policy discussion once the proposal left the world of theory and entered the political realm.

Instead, the final set of allocation rules was an amalgamation of intrinsic and instrumental principles. Focused on a series of specific allocation issues, the intrinsic and instrumental arguments shaped the debate in a manner summarized in Table 3.6. The first column of the table indicates how the original bill adopted a strong intrinsic approach to allocation favoring high-emitting plants. Congressional proposals to modify the bill generally took an instrumental approach, creating rules that further modified the prior use standard. The process thus followed the path of a Hegelian allocation presented in Chapter 2: starting with a predominantly intrinsic position, the bill gradually added instrumental modifications that still respected the original principles and were consistent with the other rules in the bill. Rather than reject the historic entitlement approach altogether, for example, legislators created growth allowances for certain states by adding a fixed percentage increase to the existing intrinsic allocation. Inconsistent equity arguments fared poorly in the process, as the Hegelian model requires. The final result, as Table 3.6 illustrates, represented an allocation outcome that was quite "Hegelian."

In addition, intrinsic arguments were most effective with respect to the question of fuel consumption by power plants, whereas instrumental norms were much more compelling with respect to the emissions rate. This pattern also fits the expected pattern of greater influence by intrinsic norms for uses seen as closer to the Lockean idea of tangible, beneficial use. Fuel consumption and power generation are much more conducive to claims based on prior use than toxic emissions, so it is not surprising that most of the movement away from intrinsic norms of allocation concerned emissions rates rather than fuel loads. This pattern will also be borne out in the grazing case, in Chapter 4, in which the particular prior use in question was even closer to the Lockean ideal, and intrinsic norms of allocation were even more influential as a result.

Another remarkable aspect of the acid rain program debate is how frequently the participants noted the "unprecedented" nature of their legislative undertaking. The administration called the acid rain program the "first major effort" to use a market-based mechanism in environmental law, and other members of Congress echoed this sentiment.[106] Senator Domenici of New Mexico was especially eloquent on this point, proclaiming that "... we have never had a situation like this, where we had to allocate these kinds of allowances, a brand new thing in environmental law."[107] Nearly everyone interviewed for this research agreed that there were few or no other models for the CAAA acid rain program while it was under consideration in Congress. In the minds of most people, creating and allocating this type of licensed property right to solve an environmental problem was a brave new world of policymaking.

In reality, Chapter 1 has already shown how the licensed property idea connects the acid rain case to numerous similar examples in the United States, dating at least to the Civil War. Although the type of right created in

Table 3.6 Changes in Allocation Rules for SO_2 Allowances

Issue	*Initial rules*	*Proposed changes*	*Final rules*
Cost-sharing for dirty plants	INTRINSIC/HEGELIAN Rate gap favoring high emitters	INSTRUMENTAL Create bonus pool for early actors; set equal rates for all emitters	HEGELIAN Bonus pool adopted; initial rate gap favoring high emitters preserved
Baseline issues	INTRINSIC Actual fuel or rate used in baseline period	INSTRUMENTAL Provide flexibility in choice of baseline year; switch to permitted emissions rates	HEGELIAN Few adjustments made to intrinsic fuel standard; shift made to permitted rates
Growth allowances for clean plants	INTRINSIC None given	INSTRUMENTAL Disregard historical fuel use patterns	HEGELIAN Allocation 20% more than historical levels
Inclusion of industry	INTRINSIC Not included	INSTRUMENTAL Polluter pays: include in program	INTRINSIC Not included
Access for new sources	INTRINSIC No redistribution to new sources: 100% offset rule	INSTRUMENTAL Weaken 100% offset requirement	INTRINSIC Market access improvements only

each example varied, all were sufficiently property-like to warrant being called a form of licensed property. All also required various allocation principles for their initial distributions. Thus, the equity questions raised by the CAAA bill in 1990 were not unprecedented or without relevant historical models, despite the perceptions of many of those closely involved in the debate.[108] In particular, the questions raised and arguments made regarding the allocation of grazing rights under the Taylor Grazing Act of 1934 are strikingly similar to the modern acid rain case. The next chapter will discuss the grazing case in detail, making those connections and comparisons with the modern case and demonstrating that important historical models for this type of policy do exist.

CHAPTER 4

ALLOCATING PUBLIC LANDS FORAGE, 1934–1938

We have a national Act known as the Taylor Act. As to what theory it was founded on, I do not know. I doubt whether anyone knows.

— Farrington Carpenter, director, Division of Grazing

Six decades before the Clean Air Act Amendments (CAAA) of 1990, Congress faced a similar set of policy questions in an entirely different ecological and political context.[1] Since the 1870s, livestock owners had pastured their sheep and cattle on the federal lands of the western United States without legal permission. For decades, the federal government largely ignored this situation, making the public range more or less an open-access resource to be exploited at will by private stockmen. Eventually, however, drought and continued overgrazing resulted in sufficient resource damage to move Congress to action. The result was the Taylor Grazing Act (TGA) of 1934, which effectively ended homesteading and created a system to control livestock grazing on the public domain for the first time.

Although federal grazing and acid rain policies are rarely mentioned in the same journal, let alone the same sentence, they have some uncanny similarities. Like the acid rain program, the TGA created a system of licensed property rights (called grazing licenses or permits)[2] to control and allocate the use of an overburdened natural resource. Because the federal range was substantially overstocked, the allocation of these rights was at least as contentious as the 1990 process. Both allocations focused on equity considerations, despite later being described as grandfathering rights or otherwise catering to powerful private interests. Both also expressed a basic tension between intrinsic and instrumental principles of ownership in reaching a Hegelian outcome.

A link between equity norms and the TGA may come as a surprise to those who have studied the public lands previously. The allocation of public domain forage has a reputation as a crass and self-interested process, dominated by powerful range users. Although the outcome did favor certain users of the public domain, the initial process was far more complicated, contested, and concerned with fairness than has been previously related. As one top grazing official noted in 1937, 98% of their administrative problems arose from determining the allocation of grazing licenses and permits.[3] From 1934 until completion of the first Federal Range Code in 1938, allocation was an open question in which the influence of conflicting norms waxed and waned. Once again, the primary axis of conflict was between intrinsic and instrumental arguments (see Figure 4.1). This chapter describes the government's struggle to allocate the public range from 1934 through 1938, concluding that equity-based norms again played a vital role in the process.

Despite the many parallels, there are of course some important differences between the two cases. Unlike the initial distribution of sulfur dioxide (SO_2) allowances, the rules for allocating grazing licenses were poorly specified by Congress. Vague statutory language led to an intense struggle by administrators and stockmen to develop specific allocation principles. Because the administrative process in this case was so important to the eventual distribution of grazing permits, this chapter devotes more attention to the early implementation of the law than to the legislative process.

In addition, although both cases created licensed property rights, the nature of the resource being allocated differed significantly. Table 4.1 makes this comparison explicit. The CAAA allocated the atmospheric capacity to absorb sulfur dioxide, whereas the TGA allocated range forage. SO_2 allowances are measured in tons, are freely transferable, and carry over into future years if not immediately used. Grazing permits are measured in animal unit months (AUMs), typically defined as the amount of forage

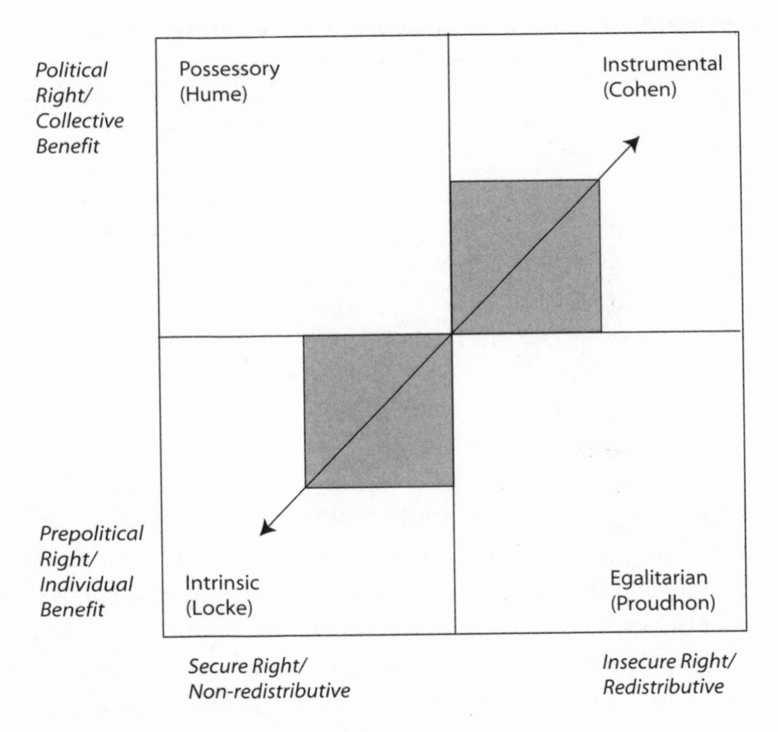

Figure 4.1 Axis of Primary Allocation Conflict: Grazing

required to feed one mature cow and one calf for one month. Forage availability and quality vary by location, so grazing permits are limited to specific parcels of land, sometimes in the form of exclusive access and sometimes in common with other users. Forage also has to be used each year; living vegetation cannot be "banked" in the same manner as sink capacity for SO_2. For these reasons, specification of the type of right in the grazing case was more complex.

Finally, the history and nature of the resource use differed between the two cases in important ways. Resource use in the SO_2 case extended back 20 to 30 years on average. Prior to the TGA, public domain grazing by individual ranches had often occurred for 60 years or more, making prior use arguments more powerful. In addition, livestock grazing is a more visible form of resource use than air pollution, especially compared with emissions from tall stacks creating ecological problems hundreds of miles

Table 4.1 Resources Allocated under the Clean Air Act Amendments and the Taylor Grazing Act

	CAAA	*TGA*
Resource	Atmospheric sink capacity for SO_2	Range forage, vegetation
Unit of measurement	Tons emitted per year	AUMs per year
Limits on time of use?	No limit; unused capacity bankable toward future years	Annual limit; unused capacity difficult to bank
Limits on place of use?	No limit; resource is consumable at any location	Significant limits; AUMs located on specific parcels of land

from the source, and is also more easily conceived of as a beneficial use of a natural resource than polluting the air we breathe. Accordingly, a ranching operation has a stronger intuitive connection to an intrinsic, Lockean principle of ownership based on sweat equity than a power plant, as was mentioned in Chapter 2. Ranchers also use the symbolic power of the cowboy in American culture to support their resource claims in ways power plant operators can only dream of. Thus, on all counts one would expect to find a stronger tilt toward a prior use standard in the TGA case than under the CAAA.

The presentation of the TGA case follows a similar pattern to the discussion of the CAAA in Chapter 3. Section I summarizes the public range history leading to the passage of the TGA in 1934. The next section considers the TGA's legislative process, including congressional hearings and the language of the law as signed. Section III describes the first two years of the law's administration (1934 through 1936), in which allocations based on intrinsic principles of prior use controlled the process despite a growing challenge from instrumental alternatives. Section IV outlines the culmination of this conflict, which resulted in allocation principles that embraced both prior use and some important instrumental modifications. Section V concludes by reviewing the parallels to the acid rain case, including the licensed property nature of a grazing permit, the prominent role of intrinsic and instrumental equity norms in the debate over distributing those rights, and the power of the Hegelian theory of ownership to describe the allocation process and outcome once more.

BACKGROUND TO THE TAYLOR GRAZING ACT

The TGA was the result of more than 50 years of conflict over the public domain. Its legal structure and allocation rules were not created *de novo*; they were modeled substantially on previous legal and extralegal approaches to controlling range access and use. To understand the TGA allocation process, therefore, it is important to understand something of this history. This section briefly reviews the pre-TGA struggles over allocation of federal range forage. In particular, the early experience of grazing administration under the Forest Service, a federal agency within the U.S. Department of Agriculture, is instructive. Many of the controversies that appeared during the first decades of the Forest Service grazing program carried over directly into regulation of the remaining public domain in the 1930s.

The Nineteenth Century: Range Wars

At the end of the Civil War, there were few livestock in the western United States. Yet by 1870, at the peak of the Texas cattle drive era, approximately 4.6 million head of cattle were in the region, and by 1890 that number had increased to 26 million (Webb 1959; Voigt 1976). Soon thereafter, sheep began to join cattle on the public lands in significant numbers as well (Barnes 1926). The late nineteenth century thereby saw an incredible boom in western livestock production fueled by the forage available on the public lands. This boom led to bitter and often violent conflict.

Public land laws of the era favored disposition to small homesteaders and made private ownership of large western pastures nearly impossible for stockmen. The result was a land tenure pattern in which ranchers acquired title to scarce riparian lands in valleys and river bottoms and used the surrounding acres of public land for seasonal pasture. Given the arid nature of much of the western range, control over rare sources of water gave these owners a strong advantage over other potential users of the public domain. Nevertheless, competition for public range forage became increasingly fierce, especially as nomadic sheep operations entered the picture. These herders eschewed private landownership, moving their flocks in annual circuits around the public lands while relying on snowmelt and other unappropriated sources of water. The resulting conflicts between nomadic herders and settled, land-owning ranchers would become a primary focus of the Taylor Act.

The federal government, committed to its policy of encouraging homesteading and subsistence farming in the West, generally declined to intervene in early grazing controversies. Instead, stockmen created extralegal institutions to allocate range rights, including illegal fencing of public lands and local claims associations organized to exclude outsiders (Dennen 1976). Within these institutions, access rights generally were based on prior use and control of adequate private property in the form of water rights—two standards that would play a crucial role decades later under the TGA. Despite these local rules, pressure from new users resulted in continued conflict and an increasingly insecure and unsatisfactory business situation for ranchers, as well as ecological damage from overgrazing (Calef 1960; Voigt 1976).

New homesteads caused more problems for the stockmen by breaking up public grazing lands with fences and private crops. Although many of these lands were poorly suited to farming, political and popular support went largely to the "nester" rather than the "cattle baron" (Peffer 1951). Thus, the nineteenth century ended with the so-called stockman's dilemma. On the one hand, cattle and sheep owners resisted any federal leasing program that potentially limited their existing local control over public land forage. On the other, they faced a growing threat from new stockmen as well as new homesteaders. By 1900, a significant rift had developed among stockmen on the issue, with cattle owners generally favoring a national leasing program for the public domain and most sheepmen rejecting any form of federal control (Peffer 1951). This split between cattle and sheep interests would continue to influence access and allocation issues into the TGA era.

The Early Twentieth Century: The Forest Service Experience

With the passage of the Forest Management Act in 1897, grazing on some public lands came under substantial federal regulation for the first time. Initially implemented by the Department of the Interior, the act authorized the regulation of livestock in the newly minted forest reserves, many of which contained large tracts of grazing land (Dana and Fairfax 1980). Despite the 1897 law, the policy for allocating grazing access on the reserves was not settled until 1902, when the department issued a circular assigning preferences in the following order (Rowley 1985):

1. stock of residents of the reserve;
2. stock of permanent ranches inside the reserve owned by residents outside the reserve;
3. stock of persons in the "immediate vicinity" of the reserve; and
4. stock of other outsiders with some "equitable claim."

Two important points emerge from this ordering. The first is that equity played an explicit part in the allocation strategy, as is most obvious by the wording in the fourth item. The second is that prior use of the range was not an explicit requirement for access. Instead, allocation was based largely on *proximity* of applicants to the public land in question. This was an example of a class-based allocation rule within the instrumental conception of ownership—the relevant class being those living on or near the forest reserve. The idea that proximity to the public lands conveyed greater entitlement to public range forage would resurface in the "home rule on the range" ideas of TGA administrators.

With the transfer of the forest reserves to the Department of Agriculture, however, and the birth of the Forest Service in 1905, prior use gained influence. Graziers already using the national forests—as the reserves would henceforth be known—were assured protection of their "priority of use" (Rowley 1985, 57). Yet the Forest Service tried to balance the interests of current range users with a preference for small users over large ones. In particular, the agency tried to ensure that new settlers with small herds would get access to the national forests, even if that meant reducing the share of forage allocated to large ranchers with priority (West 1982). Not surprisingly, the idea of redistributing range access from current users to new arrivals was a point of controversy within the Forest Service program for the next several decades.

The agency tried in a very Hegelian manner to modify priority as an allocation standard without eliminating it altogether. It enacted a maximum and a "protective" limit, both examples of a share-based instrumental allocation rule (see Table 4.2). Permittees with herds larger than the maximum were first in line for redistributive cuts in favor of new users.[4] Protective limits, by contrast, were a minimum number of AUMs below which no rancher could have his permit reduced (Rowley 1985). Protective limits also served as a cap for new entrants: their herds could never grow larger than the protective limit, whereas prior users could never dip below that level

Table 4.2 Types of Grazing Allocation Rules

INTRINSIC RULES

Priority: Favor historic users of the public range.

INSTRUMENTAL RULES

Share-Based

Protective limits: Prevent cuts of existing users' AUMs below a minimum amount.

Maximum limits: Require cuts of existing users who hold more than a certain number of AUMs, in favor of redistribution to new or existing small users.

Class-Based

Citizenship: Require applicants to be citizens of the United States, or at least to have filed intent to become so.

Commensurate property: Require applicants to hold adequate private property (generally land or water) to complement their use of the public range as part of a settled, year-round livestock operation. May or may not include leased property or purchased feed.

"Near" rule: Favor owners of commensurate property located closest to the public range.

Subsistence users: Favor range access for stock used by subsistence farmers over all other applicants.

(West 1982). Thus, protective limits were also a limited form of protection for existing users with priority over so-called new beginners.

The Forest Service also required permittees to own "commensurate" private property (Table 4.2) (Barnes 1926). Many ranchers used the public range seasonally, feeding their stock privately grown forage for part of the year. This pattern of transhumance was especially prevalent in the national forests, many of which were at high elevations and therefore unusable during the winter because of snow. Under the commensurate property requirement, range users who lacked private lands in the vicinity of the national forests were excluded from them even if they had priority of use. This favoring of private property owners closest to the public range grew from a belief that local, taxpaying land users deserved licensed property rights to public forage more than other applicants. "This is the kind of man," stated President Theodore Roosevelt in justifying this preference, "upon whom the foundations of our citizenship rest" (Rowley 1985, 62). In this way, commensurate property was a critical class-based instrumental adjustment of access rights that pushed nomadic herders off the national forests and onto the remaining public domain, where they largely stayed until the passage of the TGA (Voigt 1976).

With its share and class-based rules, early allocation policy under the Forest Service was quite redistributive in theory. A rush of homesteading on and near the national forests around 1910, however, put tremendous pressure on these instrumental rules (Roberts 1963). In the face of extensive demand for range access from new beginners and political pressure from established range users, the Forest Service eventually limited cuts on prior users in favor of new ones.[5] Nevertheless, the number of small permittees on the national forests grew dramatically in the period before World War I, resulting in more livestock than ever on the land.[6] Instrumental allocation principles, though weaker in practice than in theory, would continue to have a significant effect on the distribution of national forest grazing rights until controversy over grazing fees in 1926 led to major changes favoring existing users of the range.[7]

Getting to the TGA

Outside the national forests, grazing controls on the remaining public domain were much slower in coming. Continued interest in homesteading explained part of the delay; Congress passed several laws in the early twentieth century attempting to increase private ownership of western public lands (Dana and Fairfax 1980). Even though some of these laws were enacted ostensibly to help western ranchers, the acreages offered were still far too small and their primary effect was to make life more difficult for the public lands grazier (Peffer 1951). Numerous bills appeared in Congress during the 1920s to create a leasing system for public domain forage, but none made significant legislative progress. Range users remained divided over the proper remedy for the public lands grazing problem, or even in some cases over whether there was a problem at all.[8] Conflict also continued throughout the 1920s over whether the Forest Service or the Department of the Interior would administer any public domain grazing program (Peffer 1951). Advocates for devolution of the remaining public lands to the states also impeded a permanent regulatory arrangement.

Things changed significantly in the early 1930s. The failure of President Herbert Hoover's proposal to give the remaining public domain to the states strengthened other options.[9] Severe drought in many parts of the West exacerbated range damage from overgrazing and raised the issue's visibility in Washington, increasing pressure for legislative action (Peffer 1951). In addition, the Departments of Agriculture and Interior stopped fighting, publicly at least, over control of the public domain grazing program.[10] New examples of local-federal partnerships on the range, such as the Mizpah-Pumpkin Creek Grazing District in Montana, gained significant notice (Muhn and Stuart 1988). Finally, the threat of unilateral regulation by the executive branch spurred some reluctant western members of Congress into action.[11] As in the acid rain case, after years of stalemate the policy situation finally shifted on several fronts in favor of a legislative attempt to address the problem.

THE TGA LEGISLATIVE PROCESS

In 1932, a bill controlling public domain grazing passed the House for the first time (Peffer 1951). Named after its chief sponsor, Representative Don Colton of Utah, it contained many of the allocation principles included in the TGA two years later. Although the Colton bill died in the Senate, it was reintroduced at the beginning of the next Congress with sponsorship by Representative Edward Taylor of Colorado. As a senior member of Congress and a long-time opponent of federal leasing of the public domain, Taylor's support for the bill changed the complexion of the debate (Peffer 1951). House hearings on public domain grazing reform started in June 1933 and continued after an extended break in February and March 1934. On April 11, the House voted in favor of the bill. Senators held their own set of hearings on the proposed law and eventually approved it on June 12. On June 28, 1934, President Franklin Roosevelt signed the Taylor bill into law.[12]

The rhetoric of distributive equity was a regular feature of deliberations over the bill. Representative Burton French of Idaho, for example, proclaimed the need for "fair play" and equal opportunity for all public range users.[13] Secretary of the Interior Harold Ickes assured the House Public Lands Committee in 1934 of his intention to treat all range users, large and small, in an "absolutely" equitable manner.[14] The chief of the Forest Service went even further, observing the need for

> [T]he most equitable apportionment ... of the available grazing resources between the respective applicants therefor [*sic*], on the basis of their previous use and dependence, investments in land and range improvements, and other factors deserving of consideration ...[15]

His reference to "previous use" as well as "investments in land and range improvements" prefigured the important role of intrinsic and instrumental principles in the allocation debate.

Equity arguments emphasized the fair treatment of small ranchers. Taylor stated that the law tried specifically to protect the small owner of livestock on the public range.[16] Others were less optimistic, however, regarding the bill's eventual impact on small range users. "I know," noted one senator, "... that there is a feeling upon the part of a considerable number that this bill is being promoted by the big cattle and sheep men, and that it is going to be very injurious to the small man."[17] Concern for the "little man" was an overarching theme that would influence nearly every specific controversy. Against this background, the allocation debate comprised three familiar issues. The first was the type of right being created by the law. The second was the proper role of intrinsic, prior use principles in distributing grazing permits under the act. The third concerned alternative, instrumental allocation rules (in particular the commensurate property standard), also authorized under the law.

This section proceeds by focusing on each of those allocation issues as they were introduced and debated within the legislative process. Unlike the acid rain case (and as this chapter's epigraph indicates), Congress failed to create clear and precise rules for allocating grazing permits.[18] Instead, the Hegelian task of reconciling opposing allocation principles into a final distribution was left largely to the administrators. This section will therefore serve a more preliminary role, introducing the main allocation arguments and controversies that started during the legislative process but developed more fully during the implementation of the act.

The Type of Right Created

Like SO_2 allowances, grazing permits were clearly intended to be licensed property rights. The Colton and Taylor bills explicitly attempted to minimize ecological damage to the range by providing more stability and security of tenure for public lands ranchers.[19] Secretary of the Interior Ray

Lyman Wilbur, for example, supported the Colton bill by noting the need for more "certainty of tenure" for users of the public domain.[20] Grazing permits were the legal mechanism for providing that security by granting exclusive, ongoing access to selected stockmen. Because of the even greater focus on security for rights holders in this case, congressional attention to this issue was even more intensive than in the clean air example.

In particular, confusion prevailed over whether Congress was creating a right so secure as to be a vested, or legally guaranteed, right of private property. Taylor firmly declared permits to be a vested right under the TGA, only to backtrack in apparent perplexity under further questioning.[21] Similar uncertainty was evident among other legislators during the hearings.[22] Representative James Mott, for example, argued that the bill should not create vested rights, but remained "open to conviction or persuasion" that it in fact did so.[23] Meanwhile, Forest Service staff took issue with any mention of vested rights to graze public lands. Consider this exchange between Mott and Associate Forester E. A. Sherman:

> Mott: You say that is the policy of the Government to recognize a vested right by reason of being a user?
>
> Sherman: Not a vested right. It is a privilege allowed an individual but subject always to the public right.[24]

As originally considered in the House, the bill said little about vested rights. Section 3, which largely guided the permit allocation process, made no mention of vested rights until amended by the House Public Lands Committee. The version approved by the full House, however, included explicit protection for vested *water rights* held by public range users, adding that "and, in so far as consistent with the purposes of this Act, grazing rights similarly recognized and acknowledged, shall be adequately safeguarded."[25] With this language, the House made a messy and rather speculative equation of grazing rights with water rights, which are more commonly recognized as a legal form of private property. This ambiguous but potentially powerful statutory language intensified the discussions that followed in the Senate.

At the Senate hearings, it was again the Forest Service that led the fight against a vested right to graze the public domain. F. A. Silcox, chief of the Forest Service, attacked the amended version of Section 3. After calling the House amendment of "doubtful wisdom," Silcox cut to the chase: "The real purpose of this language is, I fear, to grant to the stockmen who are grazing lands on the public domain an estate or property interest in the particular lands which they have been accustomed to use ..."[26] Assistant Solicitor Rufus Poole of the Interior Department objected on similar grounds.[27] Both speakers recognized that the amendment's language was vague and uncertain, in intent as well as possible legal construction, and both opposed it primarily because of the threat of creating vested rights to public range access. In this respect, the goal of both federal agencies was similar to that of the Environmental Protection Agency for the acid rain bill: to create a stable right of resource access under the law, but one not legally protected from future diminution or elimination by the government.

Despite the protests, the Senate declined to remove the offending House language. Instead, it included a further provision in the same section that "... the issuance of a permit pursuant to the provisions of this Act shall not create any right, title, interest, or estate in or to the lands."[28] Thus, much like the Clean Air Act Amendments, the Taylor bill straddled the property issue. In this case, the equivocal language is explicitly in the statute, where Congress attempted to ensure the security of grazing rights and make clear that such rights are not a form of private property in the same sentence! The resulting uncertainty in this case would spawn dozens of lawsuits over the ensuing decades.[29]

Another late change to the bill complicated the issue even further. Sponsored by Nevada Senator Pat McCarran, a strong supporter of private grazing interests, this amendment changed the language regarding a permittee's right of renewal. It prevented the government from denying a permit renewal to a current range user who had complied with the rules of the agency and had pledged his or her grazing unit as collateral for a loan.[30] Given that during the Great Depression nearly all ranchers had loans on their private holdings, this provision effectively secured the tenure of any permittee who followed the basic rules of the program. While not creating a fully vested right of private ownership to the public lands or forage, the McCarran amendment took a significant step toward securing grazing permits against subsequent confiscation, particularly compared with the acid rain case.

In the end, the uncertainty over the type of right being created by Congress was reflected in the language of the statute itself. In reality, however, the right created by the bill was a clear example of licensed property—

designed to utilize many advantages of private ownership rights without creating a full-blown vested right to public domain forage. Recognition of this fact locates the TGA example within the family of market-based policies, including the acid rain program, and permits an analysis based on the property theory framework constructed in Chapter 2. Such recognition also might steer discussions away from determining whether grazing permits are property or not in favor of the more helpful question, "What type of property are they?"

Intrinsic Allocation Rules: Prior Use

Speakers during the hearings frequently supported the need to protect existing range users. Their arguments were Lockean in nature, based on an intrinsic norm of entitlement via personal labor. Despite this rhetoric, the language of the law itself provided relatively little support for prior use as an allocation standard. The result was an unclear mandate regarding the proper role of intrinsic principles in the allocation process.

Federal bureaucrats and politicians frequently expressed support for existing public domain users. Representative Taylor proclaimed on more than one occasion that the bill would respect existing "rights" on the range.[31] Interior officials testified that "the present occupant would be the preferred applicant," and Secretary Ickes himself assured nervous members of Congress that he had no intention "to run amuck" on the public domain, drive away existing stockmen, or "deprive them of rights to which they are entitled either under State laws or by customary usage."[32] Even the Forest Service chief acknowledged the basic need to recognize prior use: "If we did not recognize those preferential rights [based on priority], we would never have been able to put any rules and regulations on [range users]."[33]

Despite those official reassurances, current public range users remained skeptical. Stockmen's organizations decried the absence of explicit protection for prior users in the law itself.[34] J. B. Wilson of the Wyoming Wool Growers Association asked Congress to include a clause in the bill stating that "preference in the issuance of grazing permits shall be given to prior users of the range ..."[35] In this instance, cattlemen agreed completely with their wool-growing peers, stating that they were "absolutely in favor of the fullest protection to the present users."[36] These arguments were grounded firmly in the intrinsic perspective, favoring prior users based on their hard work and investment in the public lands ranching industry to date. As New Mexico rancher Burton "Cap" Mossman observed in his testimony, the bill needed to protect those who "have given of their lives and their fortunes in developing the water and the ranges and who know no other business than the livestock business."[37] Or consider the following rhetorical question posed by Oliver Lee, a New Mexico cattleman with 50 years experience on the public range:

> Who is entitled to preferential rights on the public domain? Is it some man that you are going to import from some other place, or some fellow who has not, up to the present time, been engaged in the livestock industry? Or is it the man who has spent a lifetime in the development of water and other improvements necessary to properly care for his livestock?[38]

The answer to the question was so obvious to Lee that he left it unstated: in all fairness, prior use must be recognized.

Concern on the part of existing range users was understandable: the original version of the bill contained essentially no language favoring prior use. Although the House amendment safeguarding existing water and grazing rights (so far as was "consistent with the purposes" of the act) attempted to provide some support for prior use, the language remained vague and the protection uncertain. Furthermore, Interior staff were sometimes less enthusiastic about the prior use standard, which served as a potential threat to the little man revered by the Roosevelt administration.[39] Secretary Ickes opened his remarks to the House committee, for example, by noting that the Supreme Court had already declared that previous use of the public domain for grazing has created "no prescriptive right to such use."[40] One of the secretary's staff members went so far as to suggest that permit reductions for current users would be forthcoming to improve range conditions.[41] Although Ickes soon rescinded such direct intimations by his underlings, the incident could hardly have reassured supporters of the intrinsic perspective.

Instrumental Allocation Rules: Small Users, Citizenship, and Commensurate Property

Although they were discussed less frequently, instrumental allocation rules were actually more prominent than intrinsic ones in the text of the new grazing law. The most important of these rules was class-based, privileging the owners of commensurate property over other applicants. Other instrumental rules included a requirement that applicants be U.S. citizens and a provision favoring small homesteaders using public lands forage for domestic stock. As with prior use, discussions of these allocation rules during the hearings were largely in terms of fairness and equity but drew on instrumental rather than intrinsic notions of ownership.

In general, share-based rules (especially those favoring new beginners) were not given much consideration. Given the extended period of prior use on the public domain by 1934, and the demise of Forest Service efforts at redistribution in the 1920s, stiff resistance to such share-based rules was to be expected. Ranchers still chafed at the memory of cuts by the Forest Service to comply with maximum and protective limits (see Table 4.2), although such cuts had become uncommon by the 1930s. Indeed, stockmen specifically referred to the redistributive policies of the Forest Service as a model to be avoided in the TGA case on fairness grounds.[42] The effective end of homesteading on the public lands under the law also reduced pressure from new beginners, who were now likely to be scarce.

Despite the general lack of support for redistributing AUMs, there were a few important exceptions. Taylor himself mentioned the need for redistribution, commenting that "it will take some time to readjust the grazing rights" on the public lands.[43] While not describing how such readjustments were to take place, Taylor echoed the Interior Department in hinting at a future in which grazing rights would be reduced and potentially reallocated. In addition, Section 5 of the law mandated that the secretary permit the free use of the public lands for feeding small numbers of livestock used for domestic purposes, regardless of priority or other factors. By privileging range use for small herds of domestic stock, Congress mandated a significant exception to any prior use or other rules created under the act.

The law also restricted allocation of range forage to citizens, or "those who have filed the necessary declarations of intention to become such ..."[44] Given the widespread and ethnically tinged resentment of nomadic sheep operations, frequently tended by Basque or Greek shepherds, the appearance of this particular class-based rule is not surprising.[45] Nevertheless, the requirement for citizenship was significant in that it had nothing to do with prior use and intrinsic rights; it was another instrumental modification of the standards by which these licensed property rights would be allocated.

Far more significant was the requirement under the law that applicants for grazing permits have sufficient commensurate property. Entirely absent from the original bill, the idea was added by the House as follows:

> ... preferences shall be given occupants and settlers on land within or near a district to such range privileges as may be needed to permit proper use of lands occupied by them.[46]

This was a strong potential challenge to prior users, many of whom lacked adequate commensurate property "within or near a district." Eventually, the Senate modified the language, changing "occupants and settlers" to include "landowners engaged in the livestock business, bona fide occupants or settlers, or owners of water or water rights."[47] In this way, the final version of the requirement moved even further from an intrinsic ideal by including absentee owners of land and owners of water rights (neither of whom necessarily invested much sweat equity in the private property in question) in the favored class of commensurate property owners.[48]

The commensurate property language was an important part of the congressional attack on nomadic herders. The requirement was defended on equity grounds by those drafting the TGA, as it had been by Forest Service rulemakers several decades earlier. As taxpayers and settled members of local communities, owners of private property received continuous praise from the bill's supporters.[49] Nomads by contrast paid no local property taxes and generally were not viewed as a beneficial part of the rural communities of the West. Their itinerant grazing of public lands also used by ranchers settled on private property was widely viewed as a form of theft.[50]

The wording of the commensurate property requirement threatened other prior users of the public domain besides nomads, however. The favoring of private lands located "within or near a district" reflected the local orientation of the law and made many property-owning range users nervous. Some sheep ranchers, for instance, had grazed the same public lands for decades without owning private property in the immediate area. These operators were not nomadic; they owned private ranches elsewhere. But because their private holdings were 100 miles or more from the public lands they used, they feared that they would find themselves on the wrong end of the "near rule" (see Table 4.2). This made proximity another important instrumental principle in conflict with intrinsic prior use arguments.[51]

Current range users noted with evident dismay the potential conflict between commensurate property and prior use principles. Cattleman Lee, for example, observed that the bill's commensurate property rule "rather annuls" its prior use language.[52] Lee expressed real concern that under the law, a new entrant could acquire or develop sufficient private property to then compete with existing users for access to the range.[53] Such worries were well founded: after the commensurate property requirement was enacted in spite of these objections, conflicts between commensurate property and priority became a critical part of the allocation debate during the initial administration of the statute.

Summary: Legislative Arguments

The TGA legislative process introduced the important issues in this allocation example and debated them largely on equity terms. Participants argued at length over the type of right they were creating, but the law clearly provided more than enough security to classify the grazing permit as a form of licensed property. Intrinsic arguments in favor of prior use were common, but the law itself failed to embrace the idea fully. Instrumental arguments also dot the legislative record, especially those favoring private property owners near the public range. The final law included just such a "commensurate" property standard, as well as other instrumental rules, but left their relationship to any prior use rule uncertain. Rather than explicitly balancing prior use and instrumental language into a final set of allocation principles, the TGA simply aggregated instrumental and intrinsic ideas while providing little guidance for administrators on how to synthesize them. As a result, it was only during the initial period of administering the TGA, recounted in the next two sections of the chapter, that many allocation problems were specifically addressed and resolved.

IMPLEMENTING THE LAW 1934–1936: PRIORITY IN CHARGE

During the first two years of the TGA's implementation, an extraordinary level of administrative energy was expended on the allocation issue. After the president signed the bill, Secretary Ickes created the new Division of Grazing Control (whose name was soon shortened to the Division of

Grazing). The division held a large number of public hearings and conferences in 1934 and early 1935 about setting up federal grazing districts under the law. By spring 1935, the first set of draft rules regarding range allocation and use was presented to range users. Based on these rules, local advisory boards comprising current range users were elected and the first licenses were assigned. By September 1935 the division had issued more than 14,000 grazing licenses within 30 grazing districts covering nearly the entire 80 million acres permitted under the law.[54] No sooner was this gargantuan task completed than the division began hearing appeals on the 1935 decisions and contemplating changes in the rules for the following year. In January 1936, advisory board members from every district met for the first time in Salt Lake City to discuss potential changes to the allocation rules. In March, the new rules for the 1936 season were issued, bringing the period to a close.

Colorado cattleman and attorney Farrington Carpenter became the first director of the Division of Grazing in September 1934. Carpenter's guiding vision for administering the public domain was local control. Instead of federal bureaucrats, Carpenter wanted local range users to do much of the allocation of grazing permits themselves. The key to this process would be local advisory boards, elected in each grazing district to help administer the act. Although the division would give the boards some basic rules by which to allocate permits, board members—or district advisors, as they were called—would retain considerable discretion.[55] Popular with many range users, the advisory board system also led to wide variations among districts in allocation rules, sometimes creating confusion among applicants and mistrust of the boards themselves.[56]

New employees of the Division of Grazing were largely westerners with a ranching background who shared their director's preference for local control. High-ranking members of the Interior Department, by contrast, including those in the Office of the Solicitor, were much more skeptical of Carpenter's exercise in "democracy on the range."[57] The result was frequent conflict between Division of Grazing staff and Interior Department officials in Washington. Because of these differences, this chapter will distinguish Division of Grazing staff, who worked directly for Carpenter, from Department of Interior staff, who worked in Washington for Secretary Ickes.

As it did in the legislative process, the rhetoric of fairness flourished under Carpenter's administration. Almost immediately after the law's enactment, federal spokesmen expressed their desire to eliminate injustices and be "as fair as possible" in allocating permits.[58] Carpenter was even more explicit, stating emphatically that "every effort is going to be made to see that every man with any right, who will use the slightest initiative, will have a chance to have his rights properly safeguarded."[59] President Roosevelt himself praised local board members not only for conserving the public range but also for their "effort to abolish unfair range practices."[60] Given the complexities of existing public domain customs and usage and the uncertain language of the TGA itself, achieving sound and equitable rules was a challenge. Nevertheless, the importance of getting the right rules in place early in the process was well recognized, and the resulting efforts were intense.[61] As Carpenter told a Colorado audience early in his tenure, "I have worked on these preferences until I see them in my sleep."[62]

In performing its task, the Division of Grazing focused on the same issues that arose in the legislative process: (a) defining the type of right created by a grazing permit, (b) determining the proper role for the intrinsic allocation principle of prior use, and (c) finding a corresponding place for instrumental allocation rules under the law. The rules and arguments considered were more specific, however, as the administrators and district advisors struggled to evaluate actual applications by ranchers for public domain AUMs. Both priority and commensurate property rules had strong advocates and critics during this period, with arguments on both sides of the issue still aligned along the intrinsic-instrumental axis. Initially, prior use served as the primary allocation principle, but only with the warning by administrators that commensurate property would soon assume more control. The 1936 rules that close this section mark the beginning of that shift. The pattern again resembled the first stages of a Hegelian process, starting with intrinsic allocation principles and then gradually modifying them on instrumental grounds. Only later, in the 1936–1938 period, however, would such instrumental modifications take full effect.

The Type of Right Being Created

The property-like status of the grazing permit remained a central point of discussion during the first two years of the TGA's implementation. The division generally affirmed its commitment to strong, secure rights for range users during this period. Yet the actual security of these rights was weakened by administrative innovations not found in the law itself. In addition, shortages of information about range use and commensurate property holdings also impeded the eventual allocation of more secure rights to range users. Although secure, exclusive rights to public domain forage remained the goal, the type of right assigned initially under the law was actually quite weak.

There was little doubt in the minds of Carpenter and his Division of Grazing staff that the grazing permit created by the TGA was nearly the equivalent of private property, particularly since the McCarran amendment virtually guaranteed the renewal of most permits. Publicly, Carpenter tried to downplay the importance of this provision of the law, urging his staff to use the term "grazing privileges" rather than "grazing rights."[63] In this manner, he tried to maintain the philosophical straddle on the vested versus nonvested nature of a grazing permit that first appeared during the TGA hearings. Privately, however, he was convinced that a permit, once issued, would be almost impossible to reduce or reallocate in the future.[64]

The stability and relative permanence of grazing permits made the proper and equitable initial allocation of these forage rights critical. Yet the information necessary to guide that allocation was just not available. In the acid rain case, current levels of fuel consumption and SO_2 emissions by power plants were well documented. By contrast, independent information in 1934 regarding prior use of the public domain and commensurate property ownership was essentially nonexistent. Indeed, the Division of Grazing would spend years trying to gather and verify this type of information for the purposes of range allocation. In the interim, however, the division had to control access to the public domain right away. The apparent legal stability of permits combined with the need for immediate action presented the new director with a puzzle: how to make a temporary allocation of rights until the information necessary for a more permanent distribution could be collected.

Carpenter solved this problem by splitting a legal hair. One of his first official acts was to ask Interior Solicitor Nathan Margold to approve a plan to create grazing *licenses* at first rather than permits.[65] Unlike a permit, the director argued, a grazing license would have to be renewed every year and by definition would be easily revocable. It also could be allocated by any set of rules, since the allocation principles in Section 3 of the law applied to the distribution of *permits* only. The division could then issue grazing licenses until sufficient information was available to allocate the range more permanently with permits. Apparently, the director's legal training paid off: his "persuasive argument" convinced the solicitor and the request was approved in time for the first set of grazing applications in 1935.[66] Carpenter summed up this nimble maneuvering to his staff by stating, "We are dodging permits for this year."[67] The dodge did not go unnoticed; Senator McCarran himself was reported to have been unhappy that the 1935 regulations did not offer "any real security or stability" to users.[68]

Some of the specifics of the permit-license distinction remained opaque despite its prominent role in the 1935 rules. Although a clear legal separation between the two was critical, the terms continued to be mixed up by Interior staff.[69] In addition, it was unclear at first how long the license system would remain in place. Initial reports indicated that permits would be issued as soon as the 1936 season, but Carpenter soon backed off such claims, leaving the timing of the shift an open question.[70] As the complexity of the allocation process became more apparent, the enthusiasm of the Division of Grazing for a quick transition to the more secure permits waned. Meanwhile, the director urged his staff to use the license period to reflect on the fairest rules for a more permanent allocation.[71]

There was also discussion of the other sticks in the bundle of rights created by a grazing license. Many public lands users wanted exclusive, fenced geographical allotments for their livestock on the public domain as soon as the act was passed.[72] Department of Interior staff initially expressed uncertainty about whether such fenced allotments would be available.[73] Soon it became clear that creating individual allotments for the first year would be an impossible task; licensees instead were told to graze wherever they normally did in the past.[74] This was hardly an optimal solution. As Carpenter recalled, "... we dodged the manner of range allotment by using the words 'customary use,' which in the case of many operators included a use from Honduras to Alaska."[75] Nevertheless, the division's long-term goal soon became the creation of as many fenced individual allotments as possible, to make the grazing permit a more secure licensed property right.[76]

The other important ownership power in question was the ability to transfer permits. On the national forests, grazing permits included a *de facto* power of alienation, despite Forest Service misgivings about this practice. The Division of Grazing had no such concerns; it was quite willing to see permits transferred among stockmen. What did trouble the division was the Forest Service permittees' practice of buying and selling permits along with the *livestock* rather than with the *commensurate property*. The Division of Grazing interpretation of the TGA was that private property owners should be privileged over all other claimants to use the public domain, and the rules for implementing the act should reflect that goal. Selling a permit along with the livestock to an operator lacking private property could undermine that principle. As Carpenter noted in one public meeting: "You have in northern Arizona [private] land the owners of which sold their forest rights years ago [along with livestock], and they apparently will never get them back. *We believe that is injustice to the [private] lands*."[77] For this reason, the division decided that permits would be transferable only as part of a sale of the private lands to which they were attached[78]—even though banks and other lenders urged the division to make the permits more freely transferable, as they were on the national forests.[79]

There were conflicting messages, therefore, regarding the type of right created under the law during the 1934–1936 period. In general, it was accepted within the division that the goal of the law was to create secure and exclusive rights for public domain users. Fears about unjust allocations based on incomplete information, however, led to an initial system with much less security or exclusivity. Users of the public domain in 1935 received licenses, which were freely revocable, rather than permits, which were not. Exclusive, individual allotments were also deferred, and the transfer of licenses was permitted only as part of a transfer of private commensurate property. Thus, the right initially created by the Division of Grazing was a weak version of licensed property. In the years that followed, the rights being allocated would become much stronger.

Intrinsic Allocation Rules: Prior Use

Priority of use was the dominant principle in the first allocation of grazing licenses in 1935. During initial meetings with stockmen after the TGA was passed, politicians and Interior representatives spoke favorably of a prior use standard.[80] Present users of the range argued the point even more strongly; as one declared to his peers, "Stock men, you must fight for your priority rights, this is your 'last round up.'"[81] Another rancher reminded the government officials of the fairness of the prior use standard: "Our forefathers recognized priority. Our Oregon State laws recognized priority use. I don't see where there is any justice in deviating from that policy."[82] Or as yet another rancher argued, "I don't think you can get an equitable adjustment of the use of the public domain without you [*sic*] consider priorities."[83] Supporters of intrinsic property principles also condemned the commensurate property idea as being unfair to prior users and objected to the idea of public forage entitlements for new beginners.[84] During a district advisors' conference in January 1936, for example, one speaker expressed this position vigorously:

> [T]o give them [new operators with commensurate property] now the opportunity to come in and demand a right when they have had nothing in the livestock business only commensurate property ... to come in and give them the preference right over those men back in another county who have built up a set-up, I feel would be very unjust.[85]

In this way, the basic intrinsic-instrumental clash regarding allocation came down largely to a conflict between priority and commensurate property, just as cattleman Oliver Lee had predicted during the TGA hearings.

Indeed, the conflict between priority and commensurate property was recognized from the outset as a "very deep and controversial matter."[86] The 1935 allocation rules recognized both principles on paper, granting grazing licenses to applicants in the following order:

1. those with commensurate property and prior use;
2. those with priority but inadequate commensurate property; and finally
3. those with commensurate property but no priority (see Table 4.3).[87]

In practice, however, the intrinsic principle of prior use dominated the process; licenses were issued to many applicants with insufficient commensurate property. Both Carpenter and Interior staff in Washington were convinced that the ranking of those with priority over those with commensurate property was not consistent with Section 3 of the act.[88] Nevertheless, they justified the 1935 rules using Carpenter's dodge, saying that Section 3 controlled the allocation of permits only, not licenses being issued as a temporary measure.

The decision to allocate in 1935 based on priority rather than commensurate property was a matter of both fairness and necessity. To be fair, Carpenter wanted to give prior users a chance to acquire necessary com-

Table 4.3 Order of Allocation Classes, 1935 Rules

Class	*Description*
Class 1	Those with both commensurate property and priority
Class 2	Those with priority only
Class 3	Those with commensurate property only

Note: Range forage was to be allocated in full to all applicants in one class before being granted to those at a lower level.

mensurate property holdings or gradually reduce their herds before the shift away from a priority standard took hold.[89] Necessity was a motivation because the division was nearly overwhelmed with the task of issuing more than 14,000 grazing licenses for the first time,[90] and resources to evaluate who had sufficient commensurate property were just not available.

Carpenter warned range users that the bias toward a prior use standard was only temporary. Those with commensurate property justifying more AUMs than they received in 1935 were promised better treatment in future years.[91] Permits issued on the basis of priority alone were stamped with a warning of a potential reduction in the future based on inadequate commensurate property.[92] Furthermore, Carpenter appeared to be aware that issuing licenses in 1935 on the basis of prior use would fail to halt ecological damage on the public domain; nevertheless, the need to create a fair process of allocation made that cost acceptable.[93] In the longer term, however, instrumental principles based on property ownership were to have a much stronger role.[94] "In the end," proclaimed the director, the property owner without priority "will get into the picture for the full amount of his commensurability."[95]

Meanwhile, a consensus began to emerge within the division that in the long term, priority could be a *secondary* principle of allocation, to distinguish between applicants with adequate commensurate property. "Commensurability only holds for as far as it holds out," said Carpenter early in 1935—after that, other standards such as priority would take over.[96] However, the future relationship of priority to commensurate property remained murky. At times, Carpenter and other division staff seemed to believe that priority would later have no allocation role at all.[97] Public statements on the subject were inconsistent, and uncertainty both within and outside the division persisted. But if the final outcome remained unsettled, the division was clearly starting the process of combining priority with commensurate property standards in a coherent manner.

The use of priority as the starting point of the allocation process is an important parallel with the clean air case. Despite statutory language supporting a commensurate property rule, the division began with a prior use approach and then modified that principle over time. This approach mirrors the Hegelian pattern in 1990, in which an allocation based on prior use was the foundation upon which significant instrumental modifications were subsequently made. The exact nature of the adjustments in this instance, and their potential "fractious holism" with intrinsic principles, would become a more heated point of conflict for the division in the coming years.

Instrumental Allocation Rules: Small Users, Citizenship, and Commensurate Property

Although priority was the most prominent allocation standard in 1934 and 1935, instrumental rules were also in evidence. Rules protecting small range users, for example, were a significant part of the initial allocation process. Although they were ultimately rejected, protective limits along the Forest Service model were contemplated (even appearing in at least one draft version of the 1935 allocation rules) and feared as a very real threat by large ranchers.[98] Soon, however, the Division of Grazing emphasis on rewarding private property owners made larger stockmen realize that substantial redistributions for new entrants and smaller operators were not likely.[99] Another instrumental rule, however, offered protection to very small range users. Allocation rules privileged subsistence applicants (those with up to 10 head of stock used for family consumption only) over all others.[100] In this respect, the subsistence rule was both class- and share-based, rewarding small users but requiring them to meet specific equity standards besides the size of their herds. Less dramatic than the protective or maximum limits of the Forest Service, the placement of small applicants at the top of the public domain pecking order was nonetheless an important equity-based attempt to allocate on principles other than prior use.

The dominant instrumental rule in the allocation process, however, remained the commensurate property requirement. Even as allocations were based on priority, division rhetoric in 1934 and 1935 praised commensurate property as an equitable standard. The vision of local landowners as the backbone of rural western communities remained the guiding principle. Carpenter invoked the image of local shopkeepers: "... a grocer has to have a store as well as a stock of goods, and a sheepman has to have a setup as well as a bunch of sheep. We are developing the livestock business into a legitimate business.[101] The director explained at length that since taxes on private land paid for schools, roads, and other public goods in the rural West, a law that took public range access away from those lands and lowered their value would be an injustice.[102] A commensurate property rule thereby helped achieve a vision of the common good, as instrumental property rules are intended to do.[103]

Supporters of commensurate property rules also attacked the priority standard directly. Some argued that past use of the range was based on unjust practices (such as intimidation and threats of violence against rivals) and therefore should not be recognized:

> I think you will agree with me that it was the hopes of those who drafted the [Taylor Grazing] Act and those who supported it that it would correct many evil practices that have developed without any regulation whatsoever.[104]

Carpenter made similar although less inflammatory remarks. "Would you say those who had [use of the public domain] should keep it?" he asked an audience in 1934. The director then proceeded to answer his own question: "I take [it] we do not like the way it is going now. We do not believe it is good for agriculture or the livestock business."[105] This argument rejected the status quo on instrumental grounds; things could be better if the allocation were different. Along the same lines, Secretary Ickes and others defended the commensurate property rule as favoring small ranchers, protecting their allotments against larger interests that may have historically run huge herds on the public domain without sufficient private land or water rights.[106]

A complicating factor in this debate was that no one was entirely sure what kinds of property qualified as commensurate.[107] The 1935 rules defined the term simply as "property which has livestock carrying capacity to supplement the public land range."[108] That property could take a number of forms, depending on the local conditions of public range use. North of the "snow line," where much of the land was unusable for grazing in winter, commensurate property generally consisted of private land on which hay could be grown for winter feed. South of the snow line (in Arizona and

New Mexico, for example), year-round use of the public range was common and controlled instead largely by access to water sources, many of which were privately owned.

However, this simple distinction between land in the North and water in the South was hardly exhaustive. Ranchers sought to meet the commensurate property requirement with a host of alternative arrangements, including: leasing private lands, leasing state or local lands, holding grazing permits on Forest Service properties, and purchasing hay as winter feed from private farmers. Carpenter even cited the example of a rancher who claimed as his commensurate property a field of cantaloupes.[109] The variety of property offered as commensurate forced the division to clarify its instrumental principles on this issue in some detail. The specifics of such controversies consumed a tremendous amount of the staff's time and energy as they struggled for consistent and equitable rules for all these alternative arrangements.[110] In the end, however, the basic principle of supporting private landownership and settled rural communities prevailed. Applicants leasing private lands,[111] for example, or purchasing feed grown on private lands[112] generally qualified under the commensurate property requirement, whereas those leasing state lands or holding Forest Service permits to complement their seasonal use of the public domain did not.

To make things more confusing, the location of an applicant's commensurate property also mattered. As noted in Section II, the TGA specifically granted preference to those with commensurate property "within or near a grazing district." Confusion reigned among range users on this point; many thought at first that they had to own property inside a grazing district to qualify for permits.[113] Although this interpretation was quickly rejected, opinions varied as to what the statutory phrase did require. Prior use advocates made the somewhat strained argument that *near* should mean any land customarily used in conjunction with the public range.[114] Others preferred a strict geographical radius outside of which commensurate property would no longer qualify. Initially, those equating the near rule with prior use triumphed. The 1935 rules defined *near* as "close enough to be used in connection with public range in the district in usual and customary livestock operations," essentially collapsing the (instrumental) near rule into the (intrinsic) prior use standard.[115]

The 1935 rules hardly ended the controversy. In 1936, Carpenter attempted to sell the local board members on a stronger near rule. His proposal, based in part on discussions with grazing administrators in other federal agencies, was to create two classes of commensurate property, one group within or immediately contiguous to a grazing district, the other being parcels farther away.[116] Despite his efforts, however, the local boards were reluctant to favor property close to the public domain over more distant property with a history of prior use.[117] Although the instrumental principle in favor of local landowners gained some public support at this time, it remained inferior to the prior use standard.

By the end of 1935, however, momentum began to shift away from priority. A crucial step in that process occurred at the first national district advisors' conference in Salt Lake City in January 1936. The meeting, attended by more than 500 stockmen and members of the Department of the Interior, was a remarkable gathering. Cattlemen and sheepmen, who for decades had been unable to peacefully graze the same land let alone sit together in the same room, participated side by side in the proceedings. Press coverage by western papers was extensive, and Division of Grazing staff clearly considered it a watershed event.[118] For the first time, formal rules to allocate and administer the public range were considered by stockmen from all the public domain states in one room at one time. These rules would serve as the overarching principles for specific allocation standards then created by each local district.

An entertaining and charismatic public speaker, Carpenter was in top form as he chaired both days of the meeting. His ability to influence district advisors on contentious issues is evident even from the written transcripts of the proceedings. The director's first agenda item was the status of so-called Class 2 (priority only) and Class 3 (commensurate property only) applicants for the 1936 season. Having given Class 2 applicants a chance to acquire adequate commensurate property or reduce their herds, Carpenter felt it was now time to give licenses to those with commensurate property but no priority before awarding licenses to those with priority only.

Debate on the point was spirited. Sheepmen lacking adequate property holdings despaired that the change would "ruin a lot of good families."[119] Such arguments drew implicitly on the intrinsic, Lockean position in the property framework:

> [I]f you are going to eliminate those men that have fostered the livestock industry through all the depressing times we have gone through, give other men a fair opportunity that has not accepted [*sic*] the livestock industry in the

> past, tell [the newcomer] to come in and take the place of the man who has fostered it, you are going to put this man out of business ... I think you have done something dangerous.[120]

Hearing these arguments, Carpenter astutely opened the floor to those with property but no priority to make their case, which they did forcefully. "This man [in favor of priority] spoke of a number of men that would be thrown out of business," said one speaker. "That is very true ... But if we are denied the range adjacent to our ranches, we are thrown out of business and the value is entirely taken off [our] ranch."[121]

In the end, the group voted in favor of the instrumental perspective, elevating commensurate property owners without prior use above applicants with priority only (see Table 4.4).[122] In addition, Carpenter was able to formulate publicly his ideas about how prior use might serve as a secondary allocation rule for the eventual issuance of permits rather than licenses.[123] The new idea of priority as a subsidiary allocation principle on ranges with too much commensurate property was gaining momentum, even as its role as the primary allocation standard was ending.

Summary: TGA Administration, 1934–1936

The initial period of TGA administration failed to settle many of the allocation issues raised by the original statute. The division weakened the type of right being issued with its distinction between licenses and permits. At the same time, it reaffirmed the eventual goal of creating a secure licensed property right to public domain forage. For reasons of equity and necessity, prior use assumed the dominant role in allocating licenses in 1935. Instrumental rules, including those regarding commensurate property, were widely discussed but had a secondary impact on allocation decisions at the time.

Table 4.4 Order of Allocation Classes, 1936 Rules

Class	*Description*
Class 1	Those with both commensurate property and priority
Class 2	Those with commensurate property only
Class 3	Those with priority only

Note: Range forage was to be allocated in full to all applicants in one class before being granted to those at a lower level.

Nevertheless, Division of Grazing staff were convinced that both fairness and the law required eventual adherence to these instrumental ideas, and by early 1936 there were indications that change was on the way. More intensive conflict between commensurate property and prior use, as well as regarding the definition of commensurate property and clarifying the "near rule," would follow, until these ideas were eventually reconciled in the Federal Range Code of 1938.

IMPLEMENTING THE LAW 1936–1938: RECONCILING THE INTRINSIC AND THE INSTRUMENTAL

After the district advisors' conference in January 1936, the division released a new set of grazing rules to relatively little controversy. The next six months would be auspicious for the grazing administrators. That summer, Congress removed the 80 million-acre cap on land that could be administered under the TGA. At the same time, district advisors journeyed to Washington for the first time to meet with Division of Grazing and Department of Interior staff. The tone of the meeting was upbeat, and range users and division staff alike seemed largely happy with the new system of grazing controls.

Despite this period of relative calm, however, new controversies were brewing. By fall, members of the Solicitor's Office at Interior expressed increasing discontent with the division's basic emphasis on priority, as well as the local districts' widely varying definitions of prior use. As a result, the division proposed a consistent (and relatively weak) priority rule for all districts at the second annual district advisors' conference, in December 1936. This time, range users were unwilling to support the division's agenda, but Interior staff in Washington soon forced the issue. Over Carpenter's objections, Secretary Ickes declared in January 1937 that *priority* in all grazing districts would mean only one year of use prior to passage of the TGA. Since existing local rules required anywhere from six months to 10 years of prior use for applicants, the uproar among range users over the new standard was tremendous. Soon after, in a widely watched case, the secretary awarded a grazing license to a Colorado sheepman who had directly challenged the validity of his local board's priority rule. This decision, combined with the one-year priority standard, marked early 1937 as the low point for priority as an allocation principle.

In the months that followed, however, priority regained some influence. Selected districts were soon granted two-year priority rules in exception to the one-year standard. Meanwhile, the division worked intensively to create yet another, more permanent set of rules. After more contentious discussions with range users and members of the Solicitor's Office, Secretary Ickes approved the first official version of the Federal Range Code on March 16, 1938. The code gave greater recognition to priority than the 1937 standard and represented a fairly complicated synthesis of the intrinsic and instrumental allocation principles in conflict throughout the period. The 1938 range code therefore marked a significant resolution within the allocation debate. Although further allocation discussions would continue district by district into the 1940s and 1950s, the basic balance between priority and commensurate property was largely settled.

The importance of making a fair and equitable distribution of the range was evident throughout the period. As before, problems continued to revolve around three subjects: the specific powers included in a permit or license to graze the public lands, the proper role of priority in allocating those permits or licenses, and the corresponding role of instrumental principles favoring small operators and commensurate property owners. This section again addresses each of these three in turn.

The Type of Right Being Created

The 1936 grazing rules mentioned permits for the first time but reaffirmed the need for the license system, given the continued lack of information required for a fair allocation of the public range.[124] Grazing permits (or term permits, as they were frequently called) thereby remained the carrot waved before the horse: always in sight but never in reach. In 1936, the division promised to grant term permits to all districts by 1937.[125] The 1937 version of the rules, however, omitted specific dates for changing to permits. By 1938, the timeline had been scaled back to granting term permits during that year in only one grazing district in each of the 10 states affected by the law.[126] Collection of adequate data on commensurate property and priority of use proved tremendously labor intensive. Although funding for the division's range survey program grew annually and eventually drew on the help of the Civilian Conservation Corps, the work far outstripped the manpower available. Meanwhile, range users showed increasing impatience with the delays and urged the division to issue term permits as soon as possible.[127]

Despite the growing tension about the transition to permits, range users were steadily gaining other sticks in the property rights bundle under the license system. The division continued to affirm its commitment to exclusive allotments for each rancher. "The ideal administration," stated one Interior report from this period, "would be an individual allotment for each applicant for range privileges ... principally for the fact that when a man has something to which he has an exclusive right, he takes pride in it and tries to preserve its resources."[128] Individual allotments were already common in states like New Mexico, where existing (and previously illegal) fence lines served as the typical demarcations of public domain allotments under the license system.[129] Although not every grazing district would be able to provide individual allotments for users, the goal was to do so wherever conditions permitted.

Even more dramatic was the increasing liquidity of grazing privileges. Initially, the division had insisted that grazing privileges would transfer only with specific pieces of private property. But the desire to recognize prior use as a qualification possessed by individuals, rather than pieces of property, eventually led to a weakening of this position. In fall 1937, a draft of the Federal Range Code included the power to transfer grazing permits from one piece of commensurate property to another for the first time.[130] The purpose of the new rule was to unite prior users owning inferior or distant commensurate properties with private parcels that lacked adequate priority but were nearer to the district. Reaction was mixed, but a modified version of the rule was included in the final version of the code.[131]

This notion of "dancing pinks" (pink being the color of commensurate property with prior use on Division of Grazing maps) considerably increased the property status of a grazing license. Suddenly, those with qualifications based on priority could take their grazing rights with them to a new ranch. Having bought private land that was more convenient but lacked priority, they could strip their old lands of their priority value and maintain their Class 1 status (having both priority and commensurate property). The grazing license had become a much more transferable asset, and a much stronger form of ownership as a result.

As property-like as licenses were becoming, they only made the expectations for permits even higher. The idea that permits were vested private property rights remained important, both as a goal and as a *de facto* practice on the range.[132] The McCarran amendment continued to encourage such expectations among both stockmen and Division of Grazing staff.[133] Even Interior Department employees less sympathetic to ranchers shared the view that permits would be an approximation of a private property right. "When permits are finally issued ...," concluded one Interior Department attorney, "[range users] are practically obtaining a permanent right to the use of these Government lands."[134] Although the government continued to retain the ultimate legal right to withdraw or reduce licenses or permits without compensation, the growing security of the licensed property right being created was clear.

Intrinsic Allocation Rules: Prior Use

For most of 1936, conflict over priority merely simmered. The division still talked publicly about the need to distribute AUMs on the public range based on property ownership rather than simply ratifying status quo patterns of use.[135] Yet priority, still defined idiosyncratically by the various advisory boards, retained a powerful allocation role in most grazing districts. Some range users criticized the division for relying too much on priority rules, others for relying too little.[136] In general, however, most large range users' associations continued to support priority as an important and fair allocation principle under the TGA.[137]

Late in the year, however, the priority issue began to heat up. It was increasingly apparent that there were far more Class 1 applicants (those with both priority and commensurate property) in many districts than the public range could sustain. The 1936 rules, unfortunately, were silent on how to make further reductions within this class. Instead, different districts chose different methods of winnowing the field (utilizing both instrumental and intrinsic ideas), such as making an equal percentage cut on all Class 1 applicants, tightening the priority standard further, or strengthening the commensurate property rule in various ways.[138] Many districts, however, favored a stiff priority rule as the fairest method for weeding out applicants. The division clearly gave its support to this particular approach, especially when used for allocating licenses rather than permits.[139]

Conflict over priority as an allocation standard soon focused on these local rules. The most prominent controversy involved the license applications of a sheepherder in Colorado Grazing District No. 6. Joseph Livingston owned adequate commensurate property for his livestock but was repeatedly denied a license based on the local requirement of two years' prior use in the district. He argued against the legality of such a standard, arguing instead that only an equal percentage cut for all commensurate property owners was valid under the language of the TGA.[140] Livingston had the funds and the determination to pursue his case doggedly and with professional legal help. By summer 1936, the division realized that the Livingston appeal was fast becoming a test case for the use of priority as an allocation standard. As Carpenter wrote to his assistant director, Julian Terrett,

> You realize that the effect of the priority rule is going to stand or fall by this [Livingston] decision and it is of immense importance to the whole west ... We must show in this decision a complete rationalization of the priority rule.[141]

The division accordingly spent considerable time at the highest levels wrestling with the equity issues raised by the case.[142] In October 1936, Livingston's case went to the Solicitor's Office for a recommendation to the secretary, who would make the final decision.

The case raised the profile of priority rules in general among Interior officials.[143] Rules requiring up to 10 years of prior use in some districts and only six months in others strengthened the impression in Washington that Carpenter's experiment in localism was out of control. Despite some misgivings, in 1936 the division had given local boards substantial freedom to craft different priority rules, only "suggesting" a two-year priority rule to all districts.[144] By October, in light of the Livingston controversy, the division no longer viewed such flexibility so favorably. Under pressure from the Solicitor's Office, the director decided it was time to make the local priority rules more "reasonable and uniform."[145] He planned to use the upcoming second district advisors' conference as the forum for getting the boards to agree.

This district advisors meeting, held in December 1936, had a remarkably different tone than the one 11 months earlier. Stockmen were less cooperative, and Carpenter had a much harder time advancing his agenda. To demonstrate the diversity of rules across the West, a chart of all the local priority standards (summarized in Table 4.5) was printed on the back of the conference program. Gently, the director suggested to his audience that the division had not expected such a wide range of priority rules, and that fairness indicated the need for a more consistent national standard.[146] The debate that followed filled the entire morning session, with many speakers bitterly condemning the director's position. Although Carpenter tried his best to assure the delegates that he still supported priority in general as an allocation rule, many present remained unconvinced of the wisdom of his plan for a national, one-year standard. As one Idaho rancher put it,

> I know I could not sit on the board, and go home with a uniform rule, say for two years. It would mean that I would have to have a bodyguard. I couldn't face my neighbors and my associates. It would mean that we would have to cut them out and let some man come in on a two-year set-up. It wouldn't be fair.[147]

In the end, such arguments carried the day and the convention rejected Carpenter's proposal.[148] Inauspiciously, the director had lost his first major

Table 4.5 Priority Rules in Grazing Districts, 1936[149]

District	*Priority period*
Arizona, all districts	2 years
California, District 1	4 1/2 years
California, District 2	5 years
Colorado, District 1	3 years
Colorado, District 2	6 months
Colorado, other districts	2 years
Idaho, District 1	3 years
Montana, District 2	1 1/2 years
Montana, other districts	3 years
Nevada, all districts	8 years
New Mexico, all districts	6 months
Oregon, all districts	2 years
Utah, District 2	10 years
Utah, Districts 5, 7	3 years
Utah, District 6	4 years
Utah, District 8	4 1/2 years
Utah, other districts	2 years
Wyoming, District 1	3 years

public battle with the stockmen. The conference thus marked the start of a heated dispute over priority that put the division and its director through its most difficult trials to date. From this point until the passage of the range code in 1938, the conflict would be at full boil.

Even as the district advisors reaffirmed their commitment to prior use, members of the Solicitor's Office were discussing whether it should play any role in the allocation rules at all.[150] In January 1937, the solicitor backed away from eliminating priority outright, declaring it a "most reasonable" standard for granting grazing privileges where ownership of commensurate property created too many qualified applicants.[151] However, the solicitor went on to recommend that all applicants with a single year of prior use between 1929 and 1934 be equally qualified for licenses and permits. Shortly thereafter, a national, one-year priority rule became part of the 1937 rules. As expected, many range users were unhappy with this weakening of an important intrinsic allocation rule.[152] Priority rules of two or more years were the norm in most districts, and many feared that a one-year rule would qualify so many applicants as to render priority a toothless standard. Instrumental methods of winnowing applicants, such as equal percentage reductions for all users, or stronger commensurate property rules regarding location, would become the only alternatives for reducing the AUMs distributed.

A second blow to prior use came a few weeks later with the resolution of the Livingston case. After months of disagreement, the solicitors ultimately presented the secretary a recommendation in favor of Livingston accompanied by a dissenting view against the sheepman. At a decisive meeting in March 1937, Secretary Ickes informed Carpenter and the other participants in the dispute that he believed Livingston had been "manhandled" by the advisory board and expressed displeasure with the entire range allocation process to date.[153] Soon afterward, the secretary awarded Livingston a grazing license for that year and rejected the two-year priority rule of the local board.[154] The division had lost its test case. In the wake of the one-year priority rule issued in January, the Livingston victory marked the low point for prior use as an allocation standard. The shift away from the intrinsic position would go no further.

Nearly lost in the uproar of these events, however, was an important loophole in the new rule. Exceptions to the one-year standard were permitted for specific grazing districts if approved directly by the secretary.

Although the majority of districts accepted the one-year rule,[155] a few successfully exploited this option. For a handful of districts faced with an overwhelming amount of qualified commensurate property, the division gained approval to extend the priority period to two years. Interestingly, the Solicitor's Office recommended approval of this intrinsic approach to winnowing applicants in these selected districts, despite its resistance to the same idea just a few months earlier. The large amount of qualified commensurate property versus available public range seemed to be the crucial factor; equal cuts of up to 50% for every operator regardless of priority apparently were too much instrumental justice, even for those at Interior who were generally opposed to prior use.

Carpenter must have found the first three months of 1937 among the least pleasant of his four years as director. There was by now quite a bit of hostility in his relationships with members of the Solicitor's Office and the Office of the Secretary, some of whom apparently were opposing Carpenter's still-pending confirmation as director by the Senate.[156] While fighting for priority in Washington, he was condemned by stockmen—his ostensible allies—for the one-year rule despite his contention that it was the only way to include any priority standard at all.[157] By the middle of March 1937, his relationship with Ickes hit a new low. Furious over the Livingston case, the secretary ordered a *de novo* review of all grazing rules issued to date and set up a Grazing Policy Committee, including a number of Ickes loyalists, to keep a closer eye on the director.[158] Carpenter despaired of his ability to take much more of such treatment, but with the Livingston decision finally issued, the crisis began to pass.[159]

Although the *de novo* review of the rules was soon postponed because of the urgent need to begin assigning 1937 grazing licenses, the mandate for another overhaul of the allocation rules was clear. The next 12 months would see an intensive effort aimed at creating a more permanent set of range rules—the Federal Range Code of 1938. Even as the Livingston decision was finalized, the newly appointed Grazing Policy Committee began reconsidering yet again the role of priority in allocating grazing permits.[160] By May 1937, the committee had concluded both that priority remained an important allocation principle and that the one-year rule created in January lacked "sufficient cutting edge" for the task.[161] Soon, Interior representatives were publicly criticizing the one-year priority rule as well.[162]

By fall, the division was drafting and circulating new sets of rules with a stiffer priority standard: two consecutive years or three nonconsecutive years of prior use in the five years preceding passage of the TGA.[163] Feedback was predictably mixed, although most Division of Grazing staff and advisory board members preferred it to the one-year standard. By late November, draft rules with the same priority standard were presented to and approved by 99 district advisors at yet another conference in Washington.[164] The rules were then subject to more scrutiny and revision by the Solicitor's Office and finally approved by the secretary on March 16, 1938. Carpenter's sense of completion was clear: "The new rules for 1938 ... should settle many controversial questions and be the beginning of a just and permanent solution to the range problem."[165]

The final rules were not a full return to dominance for priority. Instead, they included prior use as part of an allocation strategy that still relied heavily on instrumental rules.[166] The two- or three-year standard did make priority a stronger allocation tool than it had been in 1937.[167] With the new priority rules in place, for example, Livingston largely failed to obtain any further annual grazing licenses for his sheep in Colorado Grazing District 6.[168] More powerful priority standards, however, requiring up to 10 years of use in some districts, were no longer allowed. Instead, in Hegelian fashion, the intrinsic rule of the allocation formula had taken its place alongside the instrumental, rather than replacing or being replaced by it.

Instrumental Allocation Rules: Small Users and Commensurate Property

Rules protecting small range users grew more liberal by 1938. Before, as long as an applicant's livestock were kept for subsistence use and the applicant resided "near" the grazing district in question, a license for up to 10 animals was guaranteed. In 1938, however, Interior officials asked whether these subsistence applicants should meet other, more stringent commensurate property standards (such as adequate feed for winter months in districts north of the snow line).[169] The answer reemphasized the share-based aspects of the rule, stating that any class-based requirement beyond proximity and domestic use was irrelevant. The issue resulted in an amendment to the 1938 code making the preference for the subsistence user even more explicit.[170]

Other protections for small range users also made modest gains in this period. Between 1936 and 1938, the idea of protective limits continued to crop up during public discussions of the grazing allocation rules.[171] At first, Carpenter was wary, stating early in 1937 (at the height of the Livingston conflict) that "it is impossible and unfair at this time to attempt to so redistribute the range ..."[172] By 1938, however, controversy over the division's treatment of small range users had intensified. An independent report on the subject recommended additional share-based allocation rules, including a protective and a maximum limit.[173] Under pressure, Carpenter eventually began to support a very modest protective limit.[174] The 1938 range code permitted (but did not require) such protective limits in TGA grazing districts for the first time. Although these small steps toward instrumental, share-based rules were still a far cry from those under the Forest Service in the pre-1926 era, they were an additional indicator of the movement away from priority as the primary allocation standard.

Despite the gradual changes in share-based rules, the class-based principle of commensurate property remained the most prominent alternative to priority. The idea that the TGA was essentially a land use statute promoting the orderly development of rural communities by rewarding private ownership gained currency during this period.[175] Consequently, allocations that protected the public range against transient users lacking private property continued to be favored on equity grounds. As one Division of Grazing report to the secretary in 1937 noted regarding grazing licenses,

> Although these licenses are temporary in character, they are in the main a recognition that the dependent commensurate property upon which they are based are *entitled* to some sort of grazing privilege.[176]

As noted above, the 1936 allocation rules favored applicants with commensurate property over those with priority only. Nevertheless, the precise relationship between priority and commensurate property remained uncertain and controversial. Property owners who lacked prior use agitated for even greater access to the public domain.[177] As one stockman argued, "A sound and reasonably liberal policy must be developed to allow these [new] people to start in the livestock business."[178] Others were less enthusiastic about encouraging new entrants.[179] The Division of Grazing, for its part, now seemed fairly keen to keep new livestock operators out of the picture, with Carpenter later declaring the exclusion of new operators in this period "a sane and reasonable policy."[180]

The fiercest protests by applicants with commensurate property but no priority came in New Mexico. Starting with the issuance of the 1936 rules, small stockmen in that state fought actively for greater access to public range forage. The director conducted public meetings in the area in 1936 and 1937 to address the various petitions and grievances from New Mexico "being rained on" the division.[181] Many complaints insisted the TGA should give grazing rights to small property owners near the public domain regardless of prior use.[182] Several high-profile grazing appeals from New Mexico added to the conflict. In at least some of these cases, applicants with commensurate property lacking prior use won their claims, indicating the growing power of the instrumental rules.[183] Eventually, the conflict reached the floor of Congress, where the New Mexico Senate delegation introduced an amendment to the TGA providing four sections of public domain grazing land to all applicants in the state who raised livestock for a living.[184] Although the bill failed to pass, its very existence communicated the depth of resistance to prior use as an allocation standard in at least some parts of the country during the first four years of the act's administration.

In the debate over the definition of commensurate property, the norm favoring private land ownership continued to hold sway. Applicants leasing private land, for instance, continued to meet the commensurate property requirement, whereas those leasing public lands did not.[185] The exclusion of Forest Service permits from the class of commensurate property was especially controversial but remained in the 1938 Federal Range Code.[186] The biggest change during the period was that applicants purchasing hay for winter feed, rather than growing it on their own property, no longer qualified as commensurate property owners,[187] a decision that confirms the growing preference for private owners of land over all other applicants. Equity norms favoring settled owners and residents of western communities continued to drive these commensurate property definitions, despite the protests of those who did not meet the standards.

Even more controversial was the principle favoring private land near the public range. In 1938, the near rule was rewritten to favor land "dependent by location," defined by the range code as land "within or in the immediate neighborhood of the Federal range."[188] In earlier drafts, location and priority had oscillated in their relative importance. The de-emphasis of either principle led to controversy. When location was downgraded relative to priority in a September 1937 version of the rules, for example, several Division of Grazing staff and advisory board members strongly objected. As the regional grazier for Nevada put it, "It is my belief that this [new near rule] is entirely unjust and unfair to the local stockmen."[189] The continued difficulty of putting either standard explicitly above the other was so great that at one point, Carpenter suggested rating the qualifications of a private parcel 25% according to location and 75% according to prior use associated with the property.[190] Eventually, the final version of the code defined Class 1 applicants as property owners dependent by location *and* by priority, without further distinction. Thus, in the end the conflict between location and prior use was not resolved—an example of a continued tension between intrinsic and instrumental principles that remained even after the four-year rulemaking process.

Summary: TGA Administration, 1936–1938

From 1936 until the approval of the Federal Range Code in 1938, the battle over the allocation of TGA grazing rights intensified. At the same time, the nature of the right being created became more and more property-like, granting greater security and liquidity to successful applicants in expectation of a future in which permits would nearly be vested rights. Intrinsic rules based on priority descended from their commanding initial position to bottom out in early 1937. Instrumental rules followed an inverse path, having the greatest influence over allocation just two years after being almost irrelevant to the process. Starting with the reversal of priority and commensurate property as Class 2 and Class 3 in 1936, and culminating with the one-year priority rule and the Livingston decision in early 1937, instrumental rules significantly eroded the influence of intrinsic ones. The 1938 Federal Range Code then tried to reconcile the two perspectives more permanently in a Hegelian manner, including a stronger priority rule while retaining significant instrumental qualifications. Though hardly the final word on allocation of grazing rights, the code was an important resolution of this intense, initial period of public forage distribution.[191]

CONCLUSION

In this historical case of allocating licensed property rights to a public resource, definition of the type of right being created was a prominent issue. Federal grazing permits and licenses under the Taylor Grazing Act were not fully vested property rights, but from 1934 to 1938 they grew to include (and still retain) many of the powers incident to private ownership. More secure but less tradable than the SO_2 allowances created in 1990, grazing licenses and permits nevertheless represent a similar market-based policy instrument that relies on the creation of private rights to limit use of a natural resource. They are another good example of the idea of licensed property.

Contrary to popular understanding, equity concerns and arguments were a major focus of the first years of TGA implementation. The rhetoric of equity and the concern for a just allocation of rights permeates the record in the grazing case from 1934 to 1938. Although large users in the end obtained a significant degree of control over the public range, that result was highly contested and mediated on the basis of different ideas of fairness. Allocation arguments once again fell into the intrinsic and instrumental categories of the property theory framework. A review of the first four years of public range administration leaves little doubt that participants were arguing over how to justify particular allocations in terms of equity, even if their definitions of the term varied greatly.

In addition, the process of allocation was a far cry from grandfathering rights to status quo users. Priority played an important role in allocating grazing licenses and permits, more so than in the clean air case because of the more beneficial and tangible—that is, "Lockean"—nature of the historic use in question. But many prior users, especially nomadic range users lacking private land, were excluded by the allocation rules. Instrumental ideas, including share-based rules favoring very small users and class-based rules favoring owners of property near the public range, were important to the distribution of licensed property rights as well. Equity arguments were vital to justifying both the priority and the commensurate property standards. Grandfathering is an inadequate description of the results, just as it was in the acid rain case almost 60 years later.

Instead, the final allocation was again the product of a process that mirrored the development of property rights in Hegelian theory. Given a vague legal mandate, administrators initially tended to favor intrinsic principles of prior use. This set of principles was then followed by alternatives that swung in an instrumental direction, rewarding property ownership and proximity to the public range and limiting the influence of priority as a distributive standard. Eventually, the fractious holism expected from Hegelian theory developed in the form of the 1938 range code, in which priority and commensurate property coexisted uneasily side by side in the first serious attempt to reconcile the two ideals (rather than decide between them) under the TGA program. Even more than in the acid rain case, the allocation process in the TGA fits the Hegelian model.

In sum, the TGA example is remarkably similar to the acid rain case. Both laws created a licensed property right and allocated that right according to equity principles arrayed along a spectrum of entitlements ranging from the intrinsic to the instrumental. Both created their allocation rules in a Hegelian process starting with intrinsic rights, moving toward an instrumental position, and then reconciling the two. Although participants in the acid rain program thought their actions were unprecedented, in fact there is much they would have recognized in a process that occurred on the western ranges almost 60 years earlier.

Nor is the TGA case the only possible comparison for allocations of licensed property rights. Today, international resources are the latest subjects of similar market-based policies and allocations. Portions of the "global commons," like the high seas and the atmosphere, are coming under increased pressure for stronger private rights, including forms of licensed property. The allocation of such rights in an international setting suggests new challenges and new insights for the property theory framework, as the next chapter will explore in the context of negotiations under the United Nations Framework Convention on Climate Change.

CHAPTER 5

ALLOCATING GREENHOUSE GAS EMISSIONS, 1992–1997

How can we visualize any kind of global management, in a world so highly divided between the rich and the poor, the powerful and the powerless, which does not have a basic element of economic justice and equity.

— Anil Agarwal and Sunita Narain, Global Warming in an Unequal World

Modern environmental problems know no boundaries. Synthetic pesticides are now found in the Arctic, global warming melts shelf ice in Antarctica, and refuse litters the slopes of Mount Everest, the highest point on earth. As human economic activity expands seemingly without limit, global environmental problems appear to be multiplying at a similar clip. By the late 1980s, this situation led environmentalists to declare that the present generation has ushered in the "end of nature," having left no part of the global environment unmolested (McKibben 1989). In reaction, the nations of the world have entered into a rapidly expanding set of international environmental agreements, now numbering more than 200 (Porter and Brown 1996). On topics as diverse as climate change, waste trading, biodiversity loss, and ocean fisheries, international treaties now serve as a primary policy response. In fact, although domestic environmental policy has largely stagnated in the United States since the Clean Air Act Amendments of 1990, international environmental policy is a growth industry (Kraft 2002; Andrews 1999).

One significant subset of these international agreements comprises those related to the allocation of global resources neither owned nor controlled by any state or private actor. Often, such resources are referred to collectively as the global commons, in an apparent (and somewhat misleading[1]) nod to the work of Garrett Hardin (e.g., Vogler 2000). Examples include the atmosphere's capacity to absorb carbon dioxide (a compound widely believed to contribute to global warming) or minerals in the deep ocean seabed—both "open-access" resources outside any one nation's control and subject to unlimited exploitation by all. In some cases, property-like rights to use these resources have been proposed as a policy instrument to limit environmental damage in an economically efficient manner. Such proposals resemble the licensed property models studied in Chapters 3 and 4, and in the climate change case, they draw directly on the experiences of the U.S. acid rain program.

In terms of property norms and allocation rules, international conflicts over portions of the global commons provide a distinctive and worthwhile comparison with the domestic cases already considered here. As human affluence continues to grow and natural resources controlled by individual nations are steadily exhausted, pressure on many of these remaining open-access resources will only increase. The result will likely be more allocation battles on an international scale, and quite possibly more licensed property schemes. For those interested in such policies, therefore, the international setting promises to be of growing importance in the future.

At first glance, an international allocation is a challenging test for the property-theory framework developed in this work. Any such allocation will be significantly different from the domestic cases just considered. For one thing, the relevant actors are nation-states, whereas the property theories of Chapter 2 focused on individual rights of ownership. This difference alone threatens to render the framework irrelevant. In addition, property norms are quite likely to vary among nations. The powerful influence of intrinsic and instrumental norms of ownership in the acid rain and grazing cases, for instance, seems at least partially contingent on specific aspects of U.S. history and culture. Locke's ideas in particular have had a well-recognized impact on American law and policy from the Revolution to the present, and Hegelian notions have also been surprisingly influential in U.S. political culture.[2] To the degree that other nations have different political, legal, and ethical traditions, other normative principles are likely to come into play.

An especially important contextual factor is the unique historical pattern of use of the resource in question. While broadly adhering to Hegelian allocation principles, the distributions in the acid rain and grazing cases diverged significantly based on differences in prior resource use. In general, the closer resource users came to the Lockean ideal of a tangible, beneficial use in those cases, the stronger were the intrinsic claims of entitlement. In the international setting, patterns of previous resource use vary even more widely, in some cases being altogether absent. This also is likely to have a distinctive effect on the resulting allocation.

Given these differences, the international setting is a demanding one for the application of the property theory framework. On the other hand, the voluntary nature of most internationally negotiated treaties elevates the importance of equity issues. Because any party can legally abandon an international treaty negotiation at any time, agreements generally must be perceived as equitable by all parties in order to have a realistic chance at being adopted (Albin 2000; Young 1994; Grubb and Sebenius 1992). Indeed, certain shared notions of equity may even serve as "focal points" for the negotiation process (Schelling 1960). Even in the Machiavellian world of international relations, equity norms seem to matter. The question is *which* norms are relevant in the case of negotiations over allocating open-access environmental resources.

This chapter provides a preliminary answer, based on the ongoing negotiations regarding emissions of so-called greenhouse gases (GHGs), which have been linked to the problem of global climate change. The preliminary finding is this: although the property theory framework remains relevant, the contextual and historical differences in this case seem to have shifted the debate to a different axis, with arguments running from a Humean defense of allocation based on status quo possession, to a strong egalitarian position in the tradition of Proudhon (see Figure 5.1). In this dialogue, egalitarian principles appear far more influential than in either the acid rain or the public lands grazing cases, and Lockean arguments are almost entirely absent. Rather than a Hegelian holism of intrinsic and instrumental principles, there are the makings of a different (but related) property-based conflict with its own distinctive arguments and outcomes.

To say we are witnessing an international egalitarian revolution, however, would be premature. Climate change negotiators have yet to adopt any egalitarian allocation rules; indeed, the limited emissions reductions agreed

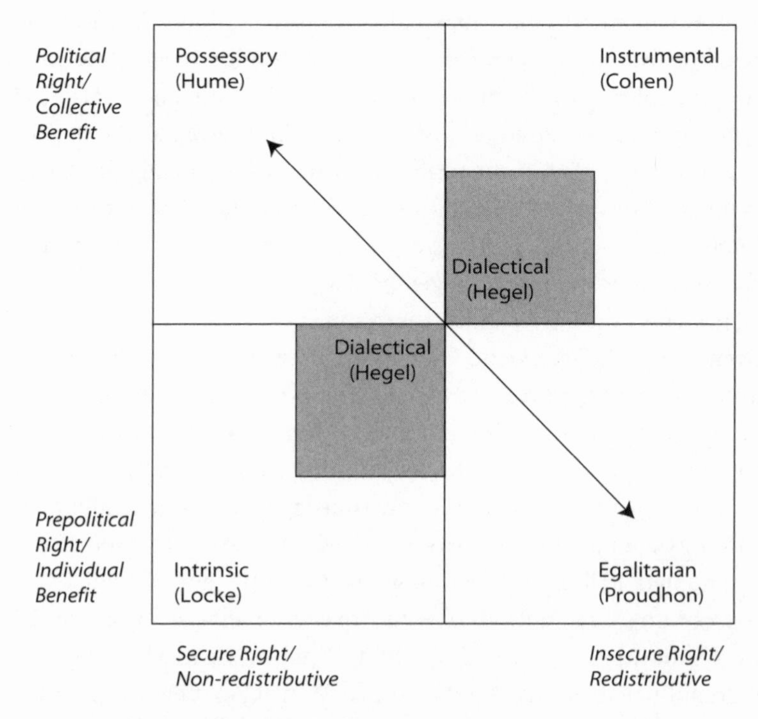

Figure 5.1 Axis of Primary Allocation Conflict: Climate Change

to so far are based mainly on the status quo ante. Given the continued resistance to egalitarian arguments by many developed nations, it is not clear just how strong egalitarian principles will be in the final analysis, especially as development pressures on this particular resource continue to increase. Instead, the climate change allocation could be moving toward a Kantian resolution, as mentioned in Chapter 2, combining egalitarian and possessory principles into a different sort of "fractious holism" than the Hegelian outcome in Chapters 3 and 4. Alternatively, it could be leading toward no resolution at all.

This chapter reviews the climate change case in more detail. It begins with a discussion of the distinctive nature of the resource: the atmosphere's capacity to absorb greenhouse gases. It then reviews the creation and ratification of the United Nations Framework Convention on Climate Change (UNFCCC), and the normative principles most evident in that process. The next section examines the first set of emissions allocations under the UNFCCC. For two years, from 1995 to 1997, nations debated the equity of various proposals for allocating GHGs, culminating in the Kyoto Protocol in December 1997. This process, though just the first step in the ongoing UNFCCC negotiations, is the best example so far of an allocation debate over GHGs and provides much to compare with the previous cases. After a brief discussion of equity issues post-Kyoto, the chapter concludes with some thoughts about how the property framework performs in this distinctive case and what future GHG allocations might look like.

It is important to note at the outset, however, that in contrast to the previous two cases, allocation of GHGs is far from complete. Consequently, any conclusions about the ultimate relevance of particular norms of ownership would be premature. But there is much to learn just from the initial, faltering steps taken by the world community toward a global allocation of GHG emissions rights in the UNFCCC process. By providing a look at this allocation in progress, the chapter will serve both as an international comparison with the U.S. cases presented in Chapters 3 and 4 and as an introduction to a promising future area of research.

BACKGROUND TO THE UNFCCC PROCESS

The problem of global climate change emerges from the overuse of one important part of the global commons: the biosphere's ability to absorb greenhouse gases. Although carbon dioxide (CO_2) and other GHGs make life possible by trapping solar radiation in the earth's atmosphere and warming the planet, in recent years there has been growing scientific concern that this beneficial "greenhouse" effect may be intensifying dangerously. Increased burning of fossil fuels, reductions in global forest cover, and other anthropogenic factors are raising atmospheric concentrations of CO_2 and other GHGs, with potentially severe long-term effects due to a warmer climate. The magnitude and consequences of these effects remain uncertain but include sea level rise and inundation of low-lying areas, more species extinctions, and increases in the frequency and scale of severe weather events (IPCC 2001).

Worries about human-induced climate change arose in international policy circles as early as 1970 (Grubb et al. 1999). Only in the past 15 years, however, has that concern led to significant political action. In 1988, mounting scientific evidence and a spate of hot summers inspired the U.N. Environment Program and the World Meteorological Organization to create the Intergovernmental Panel on Climate Change (IPCC). Composed of climate scientists worldwide, the panel was asked to synthesize and interpret peer-reviewed evidence on global warming in order to educate and inform international negotiators and policymakers. Soon thereafter, the first IPCC report confirmed the near-consensus of scientists that GHG concentrations in the atmosphere were rising because of human activity and would likely cause average global temperatures to rise as well (IPCC 1990). Subsequent IPCC reports have only intensified this concern, stating that climate change is already under way and predicting that global temperatures are likely to climb between 1.5 and 6 degrees Celsius by 2100. This represents a rate of change that is unprecedented compared with any observed fluctuations in global temperatures in the past 10,000 years (Watson 2000).

Because climate change is quintessentially a global problem—emissions anywhere on earth have the same effect on the global warming process—any policy solution will have to be international to be effective. In light of this fact, and spurred by the initial IPCC report, the United Nations created a negotiating committee that crafted the U.N. Framework Convention on Climate Change. The guiding document for international climate change policy to date, the UNFCCC was adopted by the negotiating committee and opened for signature at the United Nations Conference on the Environment and Development (UNCED) in Rio de Janiero in 1992. Subsequently, the convention was signed and ratified by a sufficient number of nations to take effect in 1994. It currently boasts more than 188 national parties.[3]

THE UNFCCC AND THE BERLIN MANDATE

With the creation of the U.N. Framework Convention on Climate Change, the international effort to control greenhouse gas emissions began in earnest. Principles of equity have figured prominently in those efforts from the outset. The framework convention itself includes a specific mandate for an equitable solution to the global warming problem (UNFCCC 1992, Article 3). It also introduced an overarching equitable principle of the "common but differentiated" responsibilities of all nations to solve the problem of climate change (Article 3). Nearly an oxymoron, this phrase indicates both the practical reality that climate change is a global problem requiring global remedies, and the ethical reality that nations have widely varying responsibility for the problem and may therefore be entitled to different shares of the atmosphere's future capacity to absorb GHGs. By recognizing this ethical principle from the start, the convention sets the stage for the allocation process in terms of norms of entitlement and fairness. Ongoing references to the parties' common but differentiated responsibilities have been a crucial part of all subsequent debates over allocating the burdens and benefits of climate change policy.

Following the framework mandate, the wider public policy community has devoted significant attention to questions of equity with respect to climate change. The Pew Center on Global Climate Change, a leading climate policy think tank, published a detailed report on equity and climate change (Pew Center 1998a) and in 2000 hosted a prominent conference on the topic. As early as 1995, the IPCC report on the economic and social aspects of climate change included two chapters devoted to equity considerations. Perhaps more tellingly, the executive summary of the volume included a discussion of fairness, with the acknowledgement that equity considerations are "an important aspect" of climate change policy (IPCC 1996, 7). Although fairness continued to take a back seat to economic considerations of efficiency in the full report, the inclusion of a detailed discussion of equity issues in such a policy-relevant document is noteworthy. Nor are these exceptional cases; many other participants in the public policy debate over climate change have also made prominent statements on equity issues (OECD 1992; UNCTD 1992; IDRC 1992).

Detailed arguments about the distribution of GHG emissions rights started with the ratification of the framework convention. The UNFCCC (1992, Article 1) introduces the allocation issue by observing that historical contributions of GHGs to the atmosphere are primarily from developed nations and that the emissions of developing nations will continue to grow. This observation is a slightly diluted version of the normative argument made by many developing nations—that countries with historically high emissions should receive fewer emissions rights in the future (Bodansky 1993; Agarwal and Narain 1991). It should be immediately apparent that this argument is a fundamental departure from the principles of the grazing and acid rain cases, in which prior use of a resource represented (at least in large part) an entitlement to, rather than a disqualification for, future access rights. This distinction is explored in more detail below.

In keeping with its emphasis on common but differentiated responsibilities, the framework convention compiles a list of nations with sizable per capita GHG emissions and greater historical responsibility for the climate change problem. This list is presented as Annex I at the end of the document and includes mainly developed nations, as well as those with "economies in transition" to a market economy, such as Russia and the nations of Eastern Europe. Drawing on this distinction, the framework sets an initial goal for Annex I countries to reduce their emissions of CO_2 and other GHGs to 1990 levels by the year 2000 (UNFCCC 1992, Article 4.2). Developing nations face no restrictions on their GHG emissions to date, although they are expected to make contributions to GHG abatement down the road (Article 4.1).

Although a laudable beginning, the framework convention's initial goal for Annex I nations was widely seen as inadequate, creating pressure for additional emissions reductions almost immediately. The idea became manifest at the first of a series of required meetings among UNFCCC signatory nations, known as Conferences of the Parties (COPs), to negotiate more detailed agreements under the framework. The first COP, held in Germany from March 28 to April 7, 1995, produced an explicit agreement known as the Berlin Mandate. Under the mandate, additional reductions were to be negotiated for the developed nations listed in Annex I, while developing nations would continue to be exempt from any quantitative emissions limits. The mandate further instructed the parties to reach agreement on these additional emissions reductions by the third Conference of the Parties, scheduled two and a half years hence. Recognizing that meetings more frequent than the annual COPs would be required to meet this deadline, the parties formed the Ad Hoc Group on the Berlin Mandate (AGBM) to work on these additional reductions. As an exhaustive, 30-month negotiation, the AGBM process remains the most detailed discussion of GHG allocations to date and thus serves as the centerpiece of this chapter.

ALLOCATION DETAILS: THE AGBM PROCESS

Over eight meetings from August 1995 to December 1997, parties to the UNFCCC met and debated in detail the merits of various allocation strategies and principles. The culmination of their work was the Kyoto Protocol, the most detailed and ambitious allocation of GHG emissions to date. The AGBM process represented only a first step toward allocating the right to emit GHGs worldwide. But it is the furthest the allocation conversation has gone in this particular setting and is instructive even as a work in progress. The record of the process contains both echoes of arguments found in previous chapters and entirely new allocation ideas. Rhetoric and arguments incorporating norms of equity and fairness were common, as they were in the grazing and acid rain cases. In particular, arguments regarding the equity of various allocation rules and strategies were prevalent but tended to fall into different categories. Here, the ideas of Hume and Proudhon took center stage, and intrinsic and instrumental allocation rules were less in evidence. Discussions over the type of right being created were also present, although in less explicit form than in the previous two examples.

The AGBM process formally began on August 21, 1995, with an initial meeting in Geneva. Representatives from 85 nations party to the UNFCCC attended, as well as those from another dozen nations as observers and several U.N. agencies and international nongovernmental organizations.[4] The chair of the AGBM, Raul Estrada-Oyuela of Argentina, noted the importance of "principles of equity" and the parties' "common but differentiated responsibilities" at the start of the first session, setting the tone for the entire process.[5] Similar reminders appeared regularly up to and including the group's final meeting, when the executive secretary reaffirmed that equity had to "remain central to the Kyoto result."[6] Specific references to the need for "equitable burden sharing" among Annex I parties punctuated the intervening dialogue, as negotiations progressed from general to specific proposals.[7] In some instances, references to equity were even made specifically in terms of an initial allocation of emissions reduction commitments.[8] In short, general appeals to the need for equity or equitable outcomes were as common here as in the previous cases considered.

The initial AGBM meeting also made it clear that the process would focus on two alternative approaches: one based on "policies and measures," and one based directly on quantitative emissions limits. The policies and measures idea focused on a potential list of actions that would be taken either voluntarily or involuntarily by various Annex I nations to reduce their emissions. This approach emphasized how nations should reduce their emissions, rather than by how much. The second approach raised the issue of specific emissions reduction targets for Annex I nations, referred to during the AGBM process as quantified emissions limitation and reduction objectives, or QELROs for short. QELROs soon became the currency of the AGBM equity debate in much the same way as emissions allowances under the Clear Air Act Amendments and animal unit months under the initial administration of the Taylor Grazing Act.

Throughout the AGBM meetings in 1995 and 1996, the parties offered numerous suggestions for policies and measures, as well as formulas and criteria for determining QELROs. Although debate over policies and measures consumed a great deal of the group's time, the appropriate role of QELROs was also a leading point of contention. Supporters of QELROs tended to eschew the rigidities of the policies and measures approach in favor of greater flexibility in meeting a particular emissions goal. By the third meeting of the group in March 1996, specific ideas for QELROs began to emerge, including a proposed flat rate reduction for all Annex I nations as well as principles for differentiating those reductions, or commitments, based on various factors. The conflict between differentiated and identical QELROs for all Annex I parties was to become an important point of conflict throughout the remainder of the process.

After a breakthrough at COP2, wherein the United States accepted the need for further legally binding commitments, discussions of QELROs shifted into a higher gear.[9] The United States linked its concession on commitments to a demand for emissions trading and other modes of flexible compliance, increasing the likelihood of a licensed property arrangement and raising issues regarding the type of right being created under any such agreement. By the sixth meeting of the AGBM in Bonn in March 1997, discussions were focusing on a flat 15% reduction from 1990 emissions levels as proposed by the European Union.[10] Other proposals, including those deviating substantially from the flat rate approach, also remained prominent in the process until the very end. Further substantive and largely off-the-record discussions occurred at the seventh and eighth meetings of the group in 1997. Finally, the AGBM process concluded during COP 3 with a late-night agreement on QELROs to be included as part of the Kyoto Protocol (named after the location of the COP 3 meeting).

Throughout the AGBM negotiations and discussions, arguments focused on the same kinds of issues found in the acid rain and grazing cases. The exact property-like nature of any potential GHG emissions right, for example, was debated frequently both inside and out of the AGBM meetings. The initial distribution of emissions rights has attracted even greater attention, with arguments again focusing on two distinctive positions within the property theory framework. The following discussion will treat each of these points separately, starting with the type-of-right issue and then moving to the question of allocation.

The Type of Right Being Created

Unlike the previous cases, there is no obvious indication in the climate change setting that the parties were particularly worried about creating vested property rights to use the atmosphere. One reason for the difference is that the threat of Fifth Amendment takings claims does not apply directly to this international allocation context (although it may apply indirectly, as certain lawsuits related to the North American Free Trade Agreement have argued; see Grieder 2001). Perhaps more importantly, the discussions under the AGBM process and elsewhere have so far been primarily about emissions reductions and burden sharing, rather than about positive rights to emit GHGs. A QELRO is in some sense an anti-emissions allowance, representing a negative reduction in permitted emissions rather than a positive right to emit GHGs. Although this distinction is purely semantic in one sense (any quantified reduction results in a quantified positive emissions entitlement as well), it does have an important conceptual implication. As in the beginning of the acid rain allocation process, the question here has been initially framed not as an allocation of valuable property rights per se, but rather as the distribution of a new regulatory burden. QELROs, in other words, are not property-like rights to emit GHGs, they are the inverse of those rights. Thus, explicit arguments about the property-like nature of a right to emit GHGs were absent during the QELROs debate.

This is not to say that scholars and policy advocates working in this area have not considered the type-of-right issue. Indeed, a number of policy institutes have tackled the question head on. One group, for instance, has already questioned the fairness of granting *permanent* rights to emit GHGs under any allocation, especially one based on current emissions levels, suggesting that such rights should be for a limited duration only (ELI 1997). Others have debated the degree to which any such emissions right should be tradable, noting the concern of some that developed nations will buy their way out of domestic emissions reductions in an "immoral" manner (Pew Center 1998b).

In fact, the issue of tradability was one type-of-right discussion that did arise in the AGBM meetings. As noted, the United States made its acceptance of any binding emissions reductions dependent on a system of emissions trading modeled on the 1990 acid rain program.[11] Joined by New Zealand and Australia, among others, the United States argued that tradable emissions rights would serve the goals of equity and efficiency as part of any climate change agreement.[12] As New Zealand concluded in its own proposal to the AGBM, the use of tradable emissions rights as part of a "least cost" solution "does not neglect equity; rather, a least cost approach improves the prospects of finding an equitable outcome acceptable to all."[13]

By contrast, many AGBM participants (in particular members of the European Union) condemned emissions trading as an unethical way for high-emissions countries like the United States to meet their QELROs by purchasing allowances from poorer nations (Grubb et al. 1999). A sense that in the interest of fairness, every developed nation must make reductions in its own domestic emissions led to arguments over strictly limiting the transfer of emissions reduction credits, as they were sometimes called.[14] This soon became known as the issue of supplementarity—that is, that any credits from emissions trading toward a nation's QELRO had to be supplemental to significant domestic actions. Many developing nations also opposed the idea of tradable emissions credits, repeatedly seeking to eliminate the idea from any agreement, at least until it was based on "equitably allocated entitlements" for all nations.[15]

Perhaps the most vehement argument against emissions trading, however, regarded the subject of "hot air." Because of their changing economic fortunes, countries from the former Soviet bloc had far lower GHG emissions in 1995–1997 than they did in 1990. Since most allocation proposals started from a baseline of 1990 emissions, these "countries with economies in transition" were looking at a large surplus of emissions rights, based not on pollution control initiatives but simply on reduced economic output. Opponents of emissions trading felt that these extra emissions rights represented a windfall and should not be subject to trade. Otherwise, the argument went, this Russian and Eastern European "hot air" would flood the market and allow other nations to easily buy their way out of their own emissions reductions, without changing the total level of Annex I emissions at all. Of course, whether one considered the surplus allowances for these countries a justified entitlement or an undeserved bonanza depended fundamentally on one's normative allocation principles. But the hot air issue largely played out in the type-of-right dialogue regarding emissions trading, rather than in the larger context of appropriate emissions allocation principles.

Also important in the type-of-right discussions was the issue of security, expressed primarily in terms of whether QELROs would be legally binding.[16] One of the first roundtable discussions on QELROs during the AGBM meetings confronted this issue directly, presenting and reviewing arguments on both sides.[17] An alternative to the notion of legally binding QELROs was the idea of "hard commitments to achieve soft targets," supported by the United Kingdom, among others.[18] Under this option, the parties would adopt mandatory policies and measures for all Annex I nations (such as eliminating national subsidies to inefficient energy technologies) without fixed emissions reduction goals. The discussion over this choice in many ways resembled the debates over the degree of security or permanence of any licensed property right created in the grazing and acid rain programs. Only here again, the issue was discussed from the opposite perspective—whether any quantitative *reductions* in emissions entitlements would be secure and measurable enough to allow for emissions trading and other property-like transactions. In the end, the parties largely settled this issue with the Geneva Declaration at the fourth AGBM meeting, making a strong commitment to legally binding QELROs as one outcome of the Berlin Mandate process.[19]

Thus, although there is not yet a clear term for licensed property rights to emit GHGs along the lines of an emissions allowance, there has clearly been much discussion over what such a right might eventually look like. Despite arguments about supplementarity and hot air, tradability survived the AGBM process and appears destined to be an important quality of any GHG emissions right. In addition, such emissions rights already entail a significant degree of measurability and security in the form of binding legal commitments to QELROs. Although it is impossible to predict where any future agreements under the UNFCCC process will go, at this writing it appears that some form of tradable and secure licensed property right to emit CO_2 and other gases will be an important part of the program.

Possessory Allocation Arguments

In this instance, the distribution question boiled down to a conflict over who should make the earliest and largest reductions in GHG emissions—developed or developing nations? The United States in particular insisted throughout the AGBM process upon "meaningful developing country participation" in GHG reductions as a condition of binding emissions targets for its own economy.[20] Developing nations, meanwhile, argued that developed nations are obligated to make the first reductions in light of their much higher historical and per capita contributions to the problem.[21]

Allocation conflicts were hardly limited within the AGBM, however, to disputes between developing and developed nations. Given the Berlin Mandate's explicit instructions to focus on emissions reductions for Annex I parties, the AGBM struggled with the equally vexing question of how to distribute QELROs among developed nations. As part of the process, the parties were invited to submit their allocation ideas and proposals for Annex I QELROs to the Secretariat. These national submissions, and the discussions they inspired, encompass a remarkable array of equity-based principles for allocating entitlements to GHG emissions. Many of these principles have yet to be incorporated in actual allocations under the UNFCCC, but one would expect proposals introduced during the AGBM process to be a common starting point of future allocation discussions. Thus, these ideas represent the best and most fully developed examples of equity principles at work in this case.

Once again, positive allocation arguments have tended to fall into two main categories (with a few important exceptions, noted below): per capita distributions versus incremental movement from the status quo (IPCC 1996; Jones and Corfee-Morlot 1992). At first glance, such arguments look like another version of the conflict between the instrumental and intrinsic positions, with developing nations urging redistribution and developed nations supporting claims based on prior use. However, this initial impression is misleading. Parties to the UNFCCC have already made arguments that are significantly different from those offered in the grazing and acid rain examples. And in reality, the conflict is far more complex than a basic disagreement between Annex I and non-Annex I countries.

To begin with, explicitly normative arguments for allocations based on prior use remain exceedingly rare in the climate change case. Few if any industrialized nations have argued that their extensive historical utilization of the global capacity to absorb GHGs entitles them to permanent ownership of the equivalent percentage of emissions rights. Indeed, the IPCC (1996, 107) observed in its discussion of the allocation question that "no one in the literature appears to advocate strict status quo as an equity principle in its own right." Nor do any of the major allocation proposals in the AGBM process rely on the notion of prior use to justify a larger number of GHG emissions rights.[22] This is a real departure from the acid rain and grazing cases. Given the prominence of such prior use arguments elsewhere, what might explain the absence of Lockean arguments in the climate change setting?

One reason was already mentioned in discussing the acid rain case. Atmospheric emissions have a weaker intuitive connection to intrinsic ownership arguments than many other kinds of resource use. Such emissions are both less tangible and less obviously beneficial than practices closer to the Lockean ideal of the yeoman farmer productively tilling the earth. These qualities certainly limited the effectiveness of intrinsic arguments in the acid rain case, as was shown in Chapter 3, and it would be reasonable to expect a similar effect in the climate change setting. But the intangible and harmful nature of the resource use in question did not fully preempt prior use arguments in the acid rain case; it only limited their influence. In the climate change process arguments based on prior use are not just weakened, they are almost entirely absent. For a more complete explanation, other reasons are required.

Like sulfur dioxide emissions, GHGs are created primarily by burning fossil fuels to supply energy. The connection between the benefits of those fuels, however, and their environmental costs is even more tenuous than in the acid rain example. Developed nations have tried to argue that high levels of energy consumption in their economies have led to substantial benefits for the entire world, such as green revolution technologies, advanced medicine, or simply more global economic growth. The persistently wide disparities in quality of life standards among nations, however, remain a substantial impediment to such arguments. Instead, high historical emissions by developed nations are more easily viewed as a self-serving harmful use of a global resource, at least from the perspective of less-developed countries. Furthermore, GHG emissions have a longer-term environmental impact than either overgrazing or sulfur dioxide emissions. Unlike SO_2, for instance, a significant percentage of GHG emissions remains in the atmosphere for decades, extending and amplifying the damage done.

Finally, the historical connection between a nation's GHG emissions and its level of poverty or wealth is certainly relevant here. As an international issue, global warming policy must confront the substantial differences in both energy use and wealth among the nations of the world. Energy use is relatively well correlated with both GHG emissions and the distribution of wealth globally (Meyer 1999). An argument in favor of the existing distribution of emissions is therefore easily interpreted as an argument in favor of the existing distribution of wealth. The close relationship of GHG emissions to gross domestic product (GDP) and other economic indices, combined with the profound gap between rich and poor nations, makes intrinsic arguments in favor of the status quo ethically challenging.

Although intrinsic, prior use arguments therefore do not appear regularly in the public debate over climate change, arguments and policies favoring the status quo are common. This apparent anomaly can be explained by reference to the other quadrant on the nonredistributive side of the property theory framework in Figure 5.1—the Humean position. Although both theories stress a strong property right, Locke's theory is based on labor resulting in beneficial *use* of the resource, whereas Hume's relies instead on ownership claims justified by current *possession*. In this sense, Chapter 2 identified the Humean, or possessory, perspective as the appropriate normative underpinning for an allocation that "grandfathers" rights to current users, regardless of other factors.

Unlike intrinsic, prior use arguments, possessory allocation principles are widespread in the climate change case. For example, allocations based on present or recent emissions levels remain a prominent policy option in many academic papers on the topic, despite the absence of ethical arguments in their favor (Burtraw and Toman 1992; Tietenberg 1992; Rose and Stevens 1993; Barrett 1992). In fact, the arguments one sees in favor of status quo emissions are distinctly Humean in nature, stressing the convenience, efficiency, or political necessity of such a starting point. Support for status quo entitlements to GHGs generally comes in the guise of pragmatic considerations, including lower regulatory costs (Tietenberg 1992) as well as raw political realities that limit other possibilities (Pew Center 1998a; Burtraw and Toman 1992). Sometimes the Humean argument is only implied, as when authors recommend 100% offset requirements for new GHG sources (Dudek and LeBlanc 1990). Either way, there is little Lockean rhetoric or talk of intrinsic rights here; there is only the cold political argument that any "realistic" allocation will have to start from the status quo. In this way, the Humean perspective assumes an influential position within the global warming debate, despite the large distributive inequalities it sustains.

Nor is the power of the Humean ideal limited to academic treatments of the climate change problem. In fact, several of the most prominent allocation proposals and arguments regarding QELROs in the AGBM were firmly located in the possessory quadrant of the property theory framework. These proposals are summarized in Table 5.1 below. In general, the strongest came from the United States; weaker versions of a possessory standard were proposed by Germany, the Alliance of Small Island States (AOSIS), and the bloc of developing nations known as the G-77 and China.

The basic nature of the Humean argument has three elements in this instance. As a theory of property based on possession, it regards the status quo levels of GHG emissions as the place to start the allocation process for all nations, regardless of current disparities. The justification for this approach is the practical realities of negotiating a mutually acceptable treaty rather than any prepolitical Lockean notions of entitlement based on pro-

Table 5.1 Possessory Allocation Proposals, AGBM

Source of proposal	*QELRO principle*	*Property theory*
United States	Meaningful participation by all nations with secure, tradable emissions rights	Strongly Humean
AOSIS, supported by Denmark and the Philippines for G-77 and China	20% Annex I reduction from 1990 levels by 2005 or 2010	Moderately Humean
Germany	15% Annex I reduction from 1990 levels by year 2010, no trading	Moderately Humean
Zaire, Peru, and the Philippines for G-77 and China	Flat percentage Annex I reduction from 1990 levels by 2005 or 2010 but with additional penalties for laggards, no trading	Moderately Humean

ductive labor. Finally, the Humean view seeks strong and secure property rights with a full complement of powers, including transferability, in order to provide the most efficient social outcomes.

Although it declined to suggest specific QELROs until the very end, the United States clearly took the strongest Humean position in the AGBM debate. On all three points just described, U.S. delegates espoused what this work calls a possessory viewpoint. Seeking outcomes "that are real and achievable," they insisted that all nations must begin making GHG reductions, not just the developed world.[23] In addition, the United States opposed the idea of differentiated commitments among Annex I parties. *Differentiation* in the AGBM meant an allocation based on specific criteria that would vary the QELROs assigned to each nation. Advocated by nations promoting alternative equity norms, the idea was opposed primarily on practical grounds. As one U.S. statement on the issue put it:

> While the United States acknowledges that clear distinctions can be (and are) drawn between different Parties and groups of Parties, we do not believe that developing a complex, formulaic approach which differentiates at an individual Party level is a viable alternative at this stage in the negotiations. To date, we have seen no formula for a differentiated approach which equitably addresses all Parties' concerns. An effort to define an acceptable differentiation scheme in this legal instrument will likely derail the negotiations, by being too divisive and time-consuming, or by disadvantaging a group of countries that might then choose not to sign or ratify.[24]

One can hardly imagine a more Humean argument against alternative allocation strategies. Finally, the chapter has already noted that the United States was a leading advocate for legally binding QELROs and emissions trading—a preliminary version of a strong and secure "property right" to emit GHGs, in other words, in accordance with the Humean model.

Other nations took a possessory approach to allocation, albeit less dramatically than the United States. In particular, many supported QELROs for Annex I nations based on flat rate reductions from status quo emissions figures. The AOSIS group started this trend by suggesting a 20% cut from 1990 emissions levels for all Annex I countries by 2005.[25] Soon this idea was supported by the G-77 and China, with the proviso that "laggard" nations missing specific reduction deadlines face additional penalties.[26] Germany also embraced the flat rate approach, recommending a 15% reduction from 1990 levels by 2010 in a proposal that was eventually embraced by the

European Union as a whole.[27] Ironically, many of these same parties to the AGBM were in frequent conflict with the United States over other details of the allocation issue. But on the need to start from existing emissions levels in any QELRO scheme, they were in full agreement.

Such arguments also echoed the Humean emphasis on practicality rather than a Lockean notion of entitlement based on prior use. Consider the defense of a flat rate reduction from the status quo offered by Germany, which begins by noting that such an allocation method has "the virtue of simplicity and practicality."[28] Although there are other ways of "approaching the concept of equity" on this issue, the Germans continue, they predict "enormous practical difficulties and obstacles" in identifying, measuring, and weighing other factors related to differentiation of commitments.[29] Or, as another European Union member state concluded, starting from the status quo may be a "crude" version of equity, but it is a practical one nonetheless.[30]

None of the proponents of a flat rate reduction were as fully Humean as the United States, however. Germany was adamantly opposed to tradable emissions rights, as were the G-77 and China, and neither was in a hurry to allocate emissions reductions to all nations, developed and developing, as the United States was. Nevertheless, on the crucial issue of QELROs, all of these parties were firmly in the Humean camp, recommending fixed, equal percentage reductions from status quo emissions for all Annex I parties in the interests of practicality and the political realities of passing and ratifying the protocol.

Instrumental and Egalitarian Allocation Arguments

Of course, global inequalities in wealth and energy consumption have also inspired a host of arguments opposed to status quo allocations. At a minimum, such arguments condemned the idea of flat rate reductions in favor of differentiation among nations based on relevant equitable criteria. As one partisan of the differentiated approach concluded,

> A stringent flat rate approach should therefore in this respect not be seen as the 'default value' in the present negotiations, but rather as being outside the mandate agreed to in Berlin. Clearly, a stringent flat rate approach for setting QELROs would not take into account ... the need for equitable and appropriate contributions by each of these Parties.[31]

Arguments like this were both widespread and influential in the AGBM, as the final, differentiated QELROs of the Kyoto Protocol attest.[32] In this respect, the Humean qualities of several major allocation proposals could not fully hold in the face of this normative challenge.

In some instances, the challenge took the form of share- and class-based instrumental rules. Policy scholars have mentioned several instrumental options, including a "Rawlsian" allocation maximizing the share of the group with the fewest emissions rights, as well as allocations based on ability-to-pay for emissions reductions or relative vulnerability to climate change impacts (Pew Center 1998a; IPCC 1996; Rose 1990). Negotiators in the AGBM echoed some of these ideas and offered a few instrumental options of their own, as summarized in Table 5.2. The ability-to-pay standard, quantified as an allocation based on per capita GDP, was suggested by several nations.[33] Other instrumental proposals tended to focus on economic efficiency in one form or another. For instance, New Zealand suggested an allocation based on equalizing the marginal costs of GHG abatement, in pursuit of a lower regulatory burden.[34] Norway, among others, recommended an allocation rewarding nations with lower carbon emissions per unit of GDP.[35] Such nations have already made the easier reductions in GHG emissions, argued the Norwegians, and so deserve smaller emissions reductions than other, more inefficient emitters. The similarities of this argument to those on behalf of "clean states" in the acid rain debate from Chapter 3 should be readily apparent.

Table 5.2 Instrumental Allocation Proposals, AGBM

Source of proposal	*QELRO principle*	*Property theory*
New Zealand	Equalizing marginal cost of abatement	Instrumental
Norway, Poland (in part), Republic of Korea (in part)	GHG emissions per unit GDP	Instrumental
Estonia, France, Spain, Norway (in part), Iceland (in part), Poland (in part), Republic of Korea (in part)	GDP per capita (ability to pay)	Instrumental

Despite the presence of some familiar instrumental arguments, however, it was a more radical and redistributive idea that emerged as a leading alternative to allocations based on status quo GHG emissions. This was the concept of equal per capita emissions rights, based on the belief that "... in a world that aspires to such lofty ideals like global justice, equity, and sustainability, this vital global common [the atmosphere] should be shared equally ..." (Agarwal and Narain 1991, 13). In a striking departure from the acid rain and grazing examples, numerous academics and advocates have proposed an allocation of emissions rights among nations based strictly on population (Baer et al. 2000; Sagar 2000; Kinzig and Kammen 1998; Meyer 1999; Grubb and Sebenius 1992). Even the IPCC (1996) has recognized an equal per capita allocation as the primary alternative to one based on the status quo.

Arguments in favor of any equal per capita position are based on equity concepts that echo the ideas of Proudhon, as summarized in Chapter 2. Not simply another instrumental alternative in pursuit of the greater good, the per capita allocation proposals assume a more immutable, prepolitical entitlement to equal shares of a global resource for every person. Indeed, in their purest form they make a far more restrictive argument than any instrumental perspective could bear—that an equal per capita allocation is the *only* ethically plausible outcome (Ott and Sachs 2000). As two leading advocates put it, we live in a world where "all human beings ought to be valued equally" and thus given equal access to this particular global resource (Agarwal and Narain 1991, 10). By emphasizing the natural right of every individual to his or her fair share, regardless of other considerations, this argument disdains the concern for other collective goals so important in the instrumental perspective.

It is interesting to speculate on why strong egalitarian arguments have surfaced in the climate change case but were noticeably absent from either the grazing or the acid rain debates. Part of the answer may again lie in the stark disparities in wealth, which make intrinsic allocation arguments harder to defend in the international context. Developing nations with per capita income levels one or more orders of magnitude below the United States are far less able to buy allowances for their future economic growth, for instance, than new power companies or ranchers on the public domain. In addition, the absence of meaningful intrinsic, prior use claims may open the "ethical space" for an alternative prepolitical notion of entitlement to play a more prominent role. Though persuasive on pragmatic terms, the Humean approach to ownership has been shown to be weaker ethically and vulnerable to normative critique (Waldron 1994). At the same time, the idea of equal shares for all has proved an intuitive "focal point" (Schelling 1960) in distributive settings where other justifications like prior use are weak or missing. Thus, the nature of the resource and the context surrounding its prior use and distribution continue to affect the relative influence of various allocation arguments.

Practically speaking, the equal per capita position includes more than one possible allocation. Its most common form involves a global cap on emissions and a distribution of the resulting emission rights to nations in direct proportion to their current (or recent) levels of population. Most such proposals involve a lengthy period of transition from current emissions levels to the ultimate egalitarian objective, although the details and timelines vary (Baer et al. 2000; Sagar 2000; Meyer 1999). Interestingly, in this particular argument, prior use or current possession of the resource in question becomes neither an entitlement nor a disqualification for future ownership rights. It is simply irrelevant to determining the ethically appropriate outcome.

This form of the equal per capita argument had a remarkably high profile in the AGBM discussions compared with the previous two cases, or even compared with other environmental treaties that were potential models for the Kyoto accord.[36] The Secretariat's review of possible criteria for differentiation refers to emissions per capita as "one of the most commonly used indicators" in the AGBM dialogue.[37] Although not part of the final AGBM agreement, the idea of differentiating commitments based on per capita emissions remained in the AGBM negotiating text well into the final stages of the process.[38] Developing nations regularly noted the large gap in per capita GHG emissions between Annex I nations and members of the G-77 and China.[39] Although not all of these arguments explicitly connected a discussion of emissions per capita to a position in favor of equal per capita allocations, many suggested movement in that direction. Table 5.3 summarizes how more than one nation recommended an allocation of emissions rights that was based substantially on egalitarian, per capita emission levels.

Like the Humean suggestions in Table 5.1, per capita proposals ranged from the moderate to the more radical. Early in the AGBM meetings, Japan submitted a proposal on QELROs that condemned flat rate approaches and recommended emissions per capita as an alternative allocation principle "on

Table 5.3 Egalitarian Allocation Proposals, AGBM

Source of proposal	*QELRO principle*	*Property theory*
Switzerland	Different QELROs for groups of nations, sorted by per capita CO_2 emissions	Moderate egalitarian
France, Spain, Norway (in part), Iceland (in part), Japan (in part), Poland (in part), Republic of Korea (in part)	Equal per capita GHG emissions	Strong egalitarian
Brazil, Iran, Poland (in part)	Historical emissions	Strong egalitarian

the premise of equity."[40] Switzerland went so far as to introduce a rule for assigning differentiated emissions reductions based on per capita emissions levels. This proposal created varying QELROs for different groups of nations, sorted by their current levels of per capita CO_2 emissions.[41] Thus, nations in a higher per capita emissions group would face higher QELROs, in what could be seen as a rough first step toward more equal per capita emissions globally.

Other nations promoted the notion of per capita emissions as a specific long-term allocation objective. France, for example, proposed four categories of nations sorted by per capita emissions. Nations in categories with higher per capita emissions faced larger emissions reductions, as in the Swiss proposal. But the French went further, arguing that "convergence" on an equal per capita global emissions level would be a logical long-term goal of the process.[42] The Netherlands, speaking on behalf of the European Union, echoed this desire for "more sophisticated methods to allocate reduction targets" in the long term, "eventually leading to convergence of emissions levels based on appropriate indicators."[43] Although not an explicit endorsement of equal per capita allocations, the use of the term convergence by the Dutch alludes to the equal per capita ideas mentioned in the French proposal and defended in the "contraction and convergence" model supported by some environmental advocates (Meyer 1999).

In the most radical version of the equal per capita position, prior use goes from being irrelevant to being an ethical liability. Under the "natural debt" idea, an equal per capita distribution of emissions rights should be based on historical as well as current levels of emissions (Smith 1996; Solomon and

Ahuja 1991). This perspective uses a simple scientific fact to justify even larger allocations to developing nations than the straight equal per capita principle: many GHGs, like CO_2, remain in the atmosphere for decades before fully dissipating. A ton of CO_2 emitted in 1950 by an industrialized nation like the United States still has a significant impact on the climate in the year 2000—according to this view it should still partially count against that nation's current per capita allocation. The natural debt argument thus turns the intrinsic idea of entitlement on its head by making prior use a *disqualification* for future GHG emissions rights. In this way, it is even more redistributive from the status quo than the basic notion of equal per capita shares.

Arguments based on historical per capita emissions were also present in the AGBM meetings, albeit not in the full-blown natural debt version. More than one developing country made explicit reference to the high historical levels of emissions of many Annex I nations as a relevant factor for determining future allocations.[44] Costa Rica, for instance, began its suggested draft of a protocol by recognizing the need to take account of "historical emissions and the specific responsibilities of the countries which have contributed to a greater extent than others to the rise in concentration of these gases. …"[45] Brazil made the more specific observation that even if annual GHG emissions of developing nations equaled those of Annex I countries by 2037 (a date that would be disputed as too distant by many), they would not equal the developed nations' cumulative impact until approximately 125 years later.[46] Thus, these nations concluded, Annex I parties remain far more responsible for the climate change problem—a fact that should be considered during future allocations.

Nor were such arguments limited to rejecting QELROs for developing nations. Brazil went on to suggest a detailed allocation proposal for all Annex I nations, based on historical as well as current levels of GHG emissions and their estimated impact on global mean surface temperature increases.[47] Although the language and calculations in the Brazilian proposal were complex, the basic idea was that countries with greater historical responsibility for GHG emissions in a previous commitment period would have to make greater emissions reductions in the following period. (The proposal also included a mechanism to fund "precautionary measures" to reduce GHG emissions in developing nations, which served as the basis of the Clean Development Mechanism in the Kyoto Protocol.) Although not a complete adoption of the natural debt approach, the Brazilian proposal is an important example of allocations for high-emitting nations based partly on their cumulative, rather than current, climate change impacts.

Reconciling Conflicting Positions: A Different Kind of "Fractious Holism"?

Although most allocation proposals during the AGBM focused on simpler principles like flat rate reductions or equal per capita shares, a few took a more complex approach. Rather than presenting entirely new ideas, however, these proposals attempted to combine and reconcile possessory, egalitarian, and even instrumental principles in distinctive and innovative ways. None of these attempts at synthesizing different allocation principles were ultimately adopted, but all represent fascinating examples of how more comprehensive GHG allocations might be crafted in the future. And while not specifically reconciling intrinsic and instrumental ideas, they do negotiate carefully between opposing equity views in a manner evocative of the Hegelian model seen in Chapters 3 and 4. If not strictly examples of Bennett's (1987) "fractious holism" at work, in other words, the proposals do seem to be attempting a synthesis of conflicting ethical views that is similar in spirit.

Japan, for example, augmented its basic proposal for a 5% flat rate reduction with some egalitarian adjustments (see Table 5.4). Under the Japanese proposal, a low-emitting nation could prorate its QELRO according to the ratio of its per capita emissions to the Annex I mean.[48] For example, a nation with per capita emissions that were only 50% of the Annex I average would be allocated a 2.5% QELRO rather than a 5% reduction under this option. The Japanese also proposed a similar pro rata adjustment for nations with emissions per unit GDP lower than the Annex I average, to reward energy efficiency . This was a simple and yet substantial modification to a basically Humean proposal, using both per capita emissions figures and the instrumental factor of emissions per unit of GDP.

Other nations attempted even more complex combinations of principles. Norway, for example, recommended a "multicriteria approach" to differentiating Annex I QELROs.[49] Three factors were critical to this allocation formula: emissions intensity per unit GDP, emissions per capita, and economic wealth as measured by GDP per capita. In the Norwegians' view, all of these factors were relevant and none were sufficient individually to deter-

Table 5.4 Synthesis Allocation Proposals, AGBM

Source of proposal	*QELRO principle*	*Property theory*
Japan	5% Annex I reduction from 1990 levels by 2010, but with egalitarian and instrumental adjustments	Mixed Humean and egalitarian
Australia	Equal per capita loss in economic welfare with strong, tradable rights	Mixed Humean and egalitarian
Iceland and Norway	Mixed formula including emissions per capita, emissions per unit GDP, GDP per capita, and share of energy from renewables	Mixed Humean and egalitarian
G-77 and China	Luxury vs. subsistence emissions	Kantian

mine the allocation. As the Norwegians indicated in one submission to the group,

> CO_2 equivalent emissions per capita and per capita GDP as indicators both reflect equity concerns, but differences in national circumstances may indicate that the two indicators could preferably be combined. For instance, if CO_2 equivalent emissions per capita were to be used as an indicator alone, countries with 'economies in transition' might have come badly off because they have a relatively high level of CO_2 emissions per capita. If used in combination with GDP per capita, however, the special situation of these Parties can also be taken into account.[50]

Ultimately, Norway's proposal combines all three factors into a weighted formula, considering a nation's emissions and GDP per capita as well as emissions per unit GDP compared with the average for all Annex I nations.[51] As under the Japanese proposal, nations with below-average figures in these areas would face proportionally smaller reductions from their present emissions levels. Thus, the Norwegian idea also starts from a Humean position ratifying the status quo but utilizes various per capita and instrumental figures to significantly reshape the final QELROs.

An even more sophisticated proposal came from the Australians, perhaps the most adamant of the AGBM parties in emphasizing the need for an equitable initial allocation of emissions rights. "Ensuring a fair and equitable outcome depends crucially on the initial distribution of commitments," they concluded. "Thus, the initial distribution of commitments must explicitly address the need for equity."[52] And although they were strong supporters of the Humean idea of tradable emissions rights, the Australians rejected a flat rate approach to assigning QELROs. Such an approach, they argued, would neglect vital differences among nations and result in "inequitable abatement tasks and unfair reductions in economic welfare for individual countries undertaking such commitments."[53] Nor would something as simple as an equal per capita allocation be adequate to meet the Australian concept of fairness; more complicated indicators were required.

In fact, the Australian approach to the allocation question was unique in that it sought to equalize nations' per capita losses in *economic welfare* due to GHG reductions. Relying on a wide array of "social welfare" indicators, the Australians proposed that QELROs be assigned such that each Annex I nation would suffer an equivalent overall reduction in welfare per person.[54] Besides being ambitiously comprehensive, the allocation strategy displays an innovative mix of Humean and egalitarian principles. On the one hand, equalizing the economic burden of QELROs on a per capita basis is distinctly egalitarian. On the other, by equalizing *reductions* in economic impact, the Australians remained faithful to the basic idea of starting the allocation process from the status quo, in a Humean manner.

All of these proposals attempt to combine egalitarian and possessory principles in a coherent manner and represent potential steps toward a reconciliation of the two views. None resemble the Kantian attempt at such a synthesis, however, described in Chapter 2. Recall that a prominent modern interpretation of Kant's theory of property suggests an allocation based on some degree of "property for all"—a guaranteed minimum distribution for everyone, combined with some additional security for property based on possession. Clearly, none of the above proposals take this particular approach. Does the Kantian perspective hold little relevance, then, for understanding the climate change case despite the obvious conflict between egalitarian and possessory principles?

Not necessarily. In fact, there has been a significant Kantian undercurrent to the allocation discussion in the climate change case, including during the AGBM process. Specifically, the notion of a universal baseline of "survival" emissions has played an important rhetorical role. More than one climate change policy scholar, for example, has embraced Henry Shue's (1993) insightful distinction between "subsistence" and "luxury" GHG emissions

(Ott and Sachs 2000; Mwandosya 1999). Several participants echoed this exact language during the AGBM process, particularly among the developing nations.[55] The argument implies that any climate change agreement must protect the right to consume energy and emit GHGs for subsistence purposes without compromise. In this respect the protection of subsistence emissions is something like an immutable and prepolitical version of the "protective limit" found in the grazing case, designed to give the highest priority to small, subsistence resource users. It also clearly resembles the Kantian notion of ownership as presented by Gillroy (2000): a basic level of property for all, with inequality based on current possession beyond that guaranteed minimum.

In the end, no specific QELRO proposal was offered based on the Kantian approach during the AGBM process. Given the focus on Annex I nations, however, it is not clear that the distinction between subsistence and luxury emissions would be terribly important at this stage of the negotiations. Only when emissions allocations extend into future commitment periods and affect the developing nations of the world more directly will the true relevance or irrelevance of the Kantian standard become apparent. Until then, the ability of the distinction between luxury and subsistence emissions to reconcile this particular allocation conflict will remain untested.

EQUITY AFTER KYOTO

The members of the AGBM finally agreed on their emissions reduction targets at the third Conference of Parties in December 1997. The results, codified in the Kyoto Protocol, required reductions in the GHG emissions of developed nations for the first time. Many details regarding implementation remained unspecified, however, and subsequent COPs concentrated on the intricacies of these arrangements. Eventually, the Kyoto process culminated with the November 2001 meeting of COP7 in Marrakech, Morocco, after which a push for final ratification of the protocol began in earnest. Emissions trading relying on licensed property rights remains an integral part of the Kyoto accord—the "keystone," in the words of one analyst (Victor 2001, 7).

Given the context of the climate change case and the prevalence of arguments favoring both instrumental and egalitarian allocations, one might have expected to see substantial deviations from the status quo in the Kyoto allocation. In practice, however, this was not the case. QELRO allocations under the protocol were fundamentally based on the pragmatic Humean standard of current possession of the resource—the most recent emissions levels of each nation. Annex I nations agreed to various percentage reductions (or increases in a few cases) from their actual 1990 emissions baseline (see Table 5.5). Most Annex I nations received something close to the average emissions reduction target of 5.2%, with only a few getting substantially different QELROs. This pattern is roughly in accordance with "symmetric" burden-sharing arrangements (like flat rate reductions) in other international environmental agreements, including the Montreal Protocol (Parson and Zeckhauser 1995, 84; see also Victor 2001). Nor were there any obvious normative explanations or allocation principles that explain the variation—why the European Union would be required to make a greater reduction, for example, than the biggest global contributor of GHG emissions, the United States.

Looking at these allocations, a skeptic might ask whether despite all the rhetoric, egalitarian principles will ever have any real influence over the allocation process. Are nations ever likely to negotiate an allocation based significantly on egalitarian equity arguments, in other words, or will they

Table 5.5 Final QELROs, Kyoto Protocol

Nation	*Kyoto Protocol QELRO**
Iceland	10% increase
Australia	8% increase
Norway	1% increase
New Zealand	No change
Russian Federation and Ukraine	No change
Croatia	5% decrease
Japan	6% decrease
Hungary and Poland	6% decrease
Canada	6% decrease
United States	7% decrease
European Union members and other Annex I nations	8% decrease

* Percentage increase or decrease from actual 1990 emissions

just "talk the talk" of convergence on equal per capita shares? In fact, the Kyoto agreement notwithstanding, the ultimate allocation of these rights remains very much in doubt. Developing nations have yet to make any quantitative commitments regarding GHG emissions at all and are adamant that developed nations must make substantial reductions first. Furthermore, even scholars with generally conservative ideas on issues like the redistribution of property have argued that climate change is different from other international environmental agreements and will require something besides equal reductions from the status quo as a viable long-term solution (e.g., Parson and Zeckhauser 1995). The allocation under the Kyoto Protocol remains only a first step, and the ultimate influence of other normative principles besides the Humean possessory position, including a Kantian distinction between subsistence and luxury emissions, very much remains an open question.

Alternatively, the continuing problems negotiating a climate change treaty may indicate just how difficult reconciling egalitarian and Humean ideas will be. Rather than a Kantian outcome or some other synthesis of these opposing views, the result could be the inability to reach any agreement at all. Developed and developing nations remain deeply divided over the ultimate direction of the UNFCCC process, as part of a larger conflict over resource distribution and international environmental policy. The United States actually withdrew from the Kyoto process in 2001, at least temporarily, amid continued complaints about lack of "meaningful" developing country participation in emissions reductions. Although ratification of the protocol continues, subsequent international meetings—from the World Summit on Sustainable Development (Rio+10) to COP8—have been hamstrung by polarization and conflict between these two camps, making many less optimistic about the eventual prospects for reconciliation. Given the current circumstances, the ability of a Kantian perspective or any other property theory to help bring these conflicting perspectives into agreement remains in doubt.

CONCLUSION

What does this discussion tell us, then, about the role of equity norms in climate change negotiations? First, the property theory framework seems equally relevant in this international setting, despite the radically different context. Arguments found throughout the AGBM process remain well categorized by the framework in Figure 5.1 and again tend to center on two opposing points of view. Although the axis of conflict between normative views is different, it still fits well within the framework of property theories and allocation perspectives presented in Chapter 2.

Second, although the framework remains relevant, the specific arguments here tend to be different than in the first two cases considered. Rather than a conflict between intrinsic and instrumental positions, what seems to be developing is a dispute between possessory and egalitarian principles. There are no appeals to ownership based on productive labor in this instance, and fewer regarding the diverse and ever-changing goals of global society. Instead, practical arguments in favor of status quo emissions levels are dueling with more radical proposals for equal per capita entitlements to emit GHGs in respect of our "common humanity." How this conflict will turn out remains uncertain, but there are reasons to think that arguments in favor of more substantial redistribution may eventually be more influential than they were in either the acid rain or grazing cases. In the end, we may see another kind of synthesis between opposing property ideas, only this time based on a normative reconciliation between Hume's ideas and those of Proudhon. Or we may see an inability to reconcile these opposing views at all.

Beyond those specific observations, the examples discussed in this chapter corroborate a more general insight into the process of allocating licensed property: context matters in determining what norms are most influential in a given allocation situation. In particular, the nature and degree of existing or historical resource use continues to play a critical and consistent role. In the climate change case, it appears that an extensive but largely negative (at least for much of the world) history of resource use combined with substantial global inequalities in per capita wealth and energy use have pushed Lockean principles out of the picture. Into this vacuum, an alternative conflict has emerged between ownership claims based on simple possession and security versus those based on a more radical, egalitarian alternative.

The larger pattern among all the cases continues to be the following: the more the existing or historical use of the resource is *intangible* and *nonbeneficial,* the less the intrinsic perspective holds sway. When that prior use remains significantly tangible and beneficial, as in the acid rain and grazing cases, the intrinsic perspective plays an important role in driving the allocation process toward an eventual Hegelian outcome. When that prior use is seen as primarily intangible and harmful, however, as it was in the climate change example, the intrinsic perspective essentially vanishes. Instead, egalitarian and Humean ideas emerge in the absence of any persuasive Lockean arguments. As a result of these distinctive ethical arguments gaining influence in the climate change case, the outcome seems likely to be quite different.

In the end, however, caution is the watchword for those drawing conclusions from the AGBM process. To date, only the possessory view has exhibited much influence over actual allocation of GHGs. Despite the bold rhetoric, neither of the more radical egalitarian alternatives—equal per capita shares and the natural debt approach—has anything close to unanimous international support, and both remain bargaining positions rather than ratified principles of international law. On the other hand, many proponents of egalitarian ideas have yet to make any binding commitments. The battle, it seems, has just been joined. Some in the AGBM process have already suggested solutions reconciling possessory and egalitarian arguments, loosely parallel to the Hegelian outcomes in the grazing and acid rain cases. But other outcomes remain quite viable as well, including a more Kantian allocation protecting "subsistence" emissions, or the absence of any further reconciliation or agreement regarding the allocation question at all. In short, where the process will go from here remains to be seen. That it will continue to offer important implications for those interested in the allocation of private rights in public resources, however, seems indisputable.

CHAPTER 6

THE IMPORTANCE OF EQUITY

It is not enough, then, for the property claimant to say simply, "it's mine," through some act or gesture; in order for the statement to have any force, some relevant world must understand the claim it makes and take that claim seriously.

— Carol M. Rose, Property and Persuasion: Essays on the History, Theory, and Rhetoric of Ownership

At its core, this book is about property claimants and the arguments they make. Power plant operators, public lands ranchers, and climate change negotiators were all asserting and contesting ownership claims over public resources, with varying degrees of success. Although the rights being claimed are weaker forms of licensed property and the process is political rather than market-driven, in the end the claims remain quite similar to those made by would-be owners of private property everywhere. In the cases considered here, however, politicians and public administrators are left with the unenviable task of deciding whose claims are valid and whose are not.

This epigraph from legal scholar Carol Rose is a good way of summing up the public allocation process that sorts out these conflicting claims. It is not enough, argues Rose, for would-be owners to assert their claims based on the arguments of their choice. In fact, only property claims that are widely recognized as legitimate throughout the larger political economic system will be effective. One key to achieving this legitimacy is that claims be based on specific, well-accepted norms of fairness regarding property rights. Many advocates and decisionmakers recognized this point in the cases studied here, and the results show that they benefited from their insight. In this way, Rose's (1994, 5–6) more general point about the need for valid property claims to be "persuasive" to those around us, as she puts it, is a fitting introduction to the conclusion of this work. Equity norms help political actors determine whose property claims are persuasive and should be recognized and whose are not.

This chapter will recapitulate the book's argument along these lines and provide an overview of its implications and findings. It will also isolate some important and promising areas for future research. It proceeds in two sections. First, the chapter summarizes the role played by equity norms in the market-based environmental policies considered here. This summary also revisits the typology of property theory first introduced in Chapter 2 in light of the empirical discussions from Chapters 3, 4, and 5. Second, the chapter summarizes the implications of this research for those creating and studying market-based policies in the future. Naïve or not, the overarching conclusion is that equity norms do matter in politics, and that scholars and practitioners alike would do well to give them more direct attention.

THE PROPERTY THEORY FRAMEWORK REVISITED

The previous chapters have demonstrated that equity arguments play an important role in environmental policy conflicts. Even in the market-based policies considered in this work, with their ostensible focus on efficiency and ecological improvement, the policy debate still gravitated toward equity concerns. In the 1930s as well as the 1990s, lawmakers, administrators, and treaty negotiators wrestled at length with these issues of fairness and conflicting entitlements to property. In doing so, they argued and deliberated within a framework of relevant norms that helped structure the process and shape the outcome.

In particular, this book applied the property framework in three policy settings: the acid rain program under the Clean Air Act Amendments of 1990, the regulation of grazing on the public domain under the Taylor Grazing Act of 1934, and the allocation of greenhouse gas emissions among nations under the United Nations Framework Convention on Climate Change in the

1990s. For a variety of theoretical and institutional reasons, analysts of market-based policies have tended to neglect historical precedents (like grazing permits) for modern, market-based policies (like the acid rain or climate change examples). But despite the perception that such policies are unique to the late twentieth century, numerous historical examples exist that are helpful for understanding both current and future policy debates. Their different ecological and historical settings notwithstanding, all three cases were shown to be examples of market-based policies creating private rights in a public resource: quantifying and securing access rights for certain users, in other words, in order to limit resource use to a more optimal level.

Policies of this type, I have argued, give rise to a kind of private property that, while falling short of a fully vested right, remains quite recognizable as a form of ownership. These rights are a licensed form of property: they have many, but not all, of the powers typically assigned to a vested property right, yet like any other legal license, they remain formally subject to government adjustment or cancellation without compensation. Although those creating such policies tend to shy away from any mention of the term property in order to stave off takings claims or for other normative reasons, the facts speak for themselves. Besides being limited in terms of public compensation, these new legal rights are explicitly designed in many ways to function as a form of private ownership.

Debate over the equity implications of these policies focused on the initial allocation of rights. Different theories of property support different allocation rules, and advocates and policymakers alike drew implicitly but extensively on those property theories to defend their preferred allocation proposals. In general, equity arguments in the grazing and acid rain cases centered on the opposing intrinsic and instrumental ideas of ownership exemplified by the property theorists John Locke and Morris Cohen, respectively. In the climate change case, different ideas have taken hold, in particular those of David Hume and Pierre-Joseph Proudhon. This is not to argue, of course, that activists, bureaucrats, treaty negotiators or members of Congress were discussing the collected works of Locke, Proudhon, or any other property theorist by name. Rather, the arguments made for various allocation alternatives relied on specific *norms* of private ownership that have been heavily influenced and shaped over the years by these leading theorists. The connection here between theory and practice, I argue, is indirect but still essential to understanding the political processes at work.

It is certainly clear that the rhetoric in all three cases relied extensively on the language of equity. Appeals to fairness are abundant in the written record, and personal interviews with key policy actors also confirmed the prevalence of equity issues behind the scenes in the acid rain case. Nor could participants construct their equity arguments in any manner that was convenient—convincing allocation arguments were limited to a few specific views of property theory that have been debated and refined for centuries. Those with extensive histories of resource use argued for distributions on Lockean grounds in the domestic setting but shifted to a more Humean argument based on current possession in the international context. Others, including those with limited prior resource use, sought modifications from the historic pattern based on instrumental arguments or even more radical egalitarian views in the international case. Actors in all three cases drew heavily on the language of equity and property-based norms of entitlement in trying to shape the final allocations.

Nor were these rhetorical arguments mere window-dressing for an outcome that was actually determined by other factors, like economic self-interest and political clout. Chapter 1 reviewed how previous models using economic self-interest to explain legislative behavior have demonstrated mixed results at best. Instead, the literature has found that ideology, including ideas of equity and fairness, is a significant determinant of legislative behavior. Evidence from Chapters 3, 4, and 5 supports this role for equity norms: the final allocation rules in all three cases largely followed certain frequently referenced normative principles. More specifically, political actors in both U.S. cases followed a similar Hegelian process in finalizing the allocation rules. Early allocation proposals by policymakers tended to favor intrinsic principles. Opponents (including policymakers, resource users, and other advocates) then challenged the proposed allocation on various instrumental grounds, causing a shift away from the standard of prior use. The shift was not complete, however; instead, each set of final allocation rules reflected an effort to reconcile or synthesize the two ideals into a "fractious holism" of conflicting opposites. Arguments that were internally inconsistent or otherwise strayed too far from these dominant normative positions were omitted from the final allocation, providing further evidence of the influence of these particular norms.

In the international context, the outcomes have largely been faithful to another set of allocation principles. Here, the distribution of quantified emissions reductions of greenhouse gases among developed nations relied heavily on a Humean principle of current resource possession based on practical considerations. Although "differentiation" among nations did play a role in the final allocation under the Kyoto Protocol, the fundamental principle was clearly a nation's entitlement to a baseline of its current emissions levels, with modest modifications based on the practical realities of negotiating a treaty acceptable to all parties. Although egalitarian principles had relatively little influence over the final Kyoto allocation, they were prominent during the negotiation process and seem likely to gain importance in the future. Developing nations, who are the leading proponents of more egalitarian allocations, have yet to assume any binding emissions reduction obligations. If and when developing nations begin to receive their own allocations of emissions rights, egalitarian ideas like "contraction and convergence" may become a much more important distributive principle.

Because the allocation of greenhouse gas emissions rights is not complete, the possessory and egalitarian arguments have yet to be reconciled. Although the Hegelian model is not directly appropriate in this context, the need for a parallel kind of "fractious holism" remains obvious from the Kyoto process. Is there another theory that holds promise for resolving the tension between these particular principles? One potential candidate is based on the work of Immanuel Kant. As interpreted by modern thinkers, Kantian property theory may hold some useful ideas for creating a synthesis between possessory and egalitarian ideas of allocation. In particular, an allocation based on the distinction between "subsistence" and "luxury" emissions, with a per capita guarantee of the former but not the latter, is a Kantian allocation principle alternative already being discussed in the climate change setting. But the effectiveness of such a Kantian approach remains substantially untested; other outcomes remain viable, including the inability to reach any sort of equitable accord at all. Whatever the eventual result, however, it is certain to have important implications for the wider application of the theoretical framework developed here.

Certain aspects of the allocation context seem to play an especially important role in determining which norms are most influential. In general, since property norms vary widely from nation to nation and culture to culture, the political economic setting appears to have an important effect. Thus, the similarity of the Hegelian outcome in the two domestic cases is almost certainly due in large part to long-standing qualities of the United States' legal and political history. By contrast, the distinctive values of other nations and cultures may help explain the contrasting set of norms that seem most important in the climate change case.

One specific feature of the allocation context, however, seems particularly relevant to all the cases considered here: the degree and nature of the prior resource use. The allocation in the grazing case relied more on intrinsic property ideas than did that in the acid rain example, and I have argued that the greater influence of Locke in the former is not surprising, given the nature of the resource use being allocated. Resource use that is generally perceived as more beneficial and more tangible—that is, more closely related to the paradigmatic Lockean example of the small farmer productively tilling the land—is more likely to gain greater recognition on intrinsic grounds. Resource use that is relatively negative and intangible, such as emissions of sulfur dioxide or certain greenhouse gases, is accordingly less likely to receive recognition based on an equitable, prior use basis. Property claims based on prior use are more persuasive for ranchers, in other words, than for power plant operators. As a result, intrinsic arguments had the most influence over the final allocation in the grazing case.

That pattern raises the interesting question of situations in which prior use is almost entirely the antithesis of the Lockean ideal or is missing altogether. Chapter 5 considered one example of this situation, noting that allocation conflicts over parts of the global commons (resources outside the sovereign control of any nation) may lack any plausible connection to the Lockean model. In the case of negotiations over greenhouse gas emissions, prior use arguments by advocates on intrinsic terms are almost entirely missing, despite the extensive histories of emissions by many developed nations. Here, a lack of convincing arguments regarding the "beneficial" nature of this prior use (at least for most of the world's population) also seems to be moving the allocation debate in a different direction, toward a conflict between Humean principles based on simple possession and more radical egalitarian alternatives echoing the thoughts of Proudhon. As noted above, although the final outcome of this particular conflict is still undetermined, the chances of a more egalitarian distribution seem stronger than in either the grazing or the acid rain cases, in part because of the nature of the prior use.

In light of these findings, the allocations of property rights in all three empirical cases can be usefully classified within the property theory framework outlined in Chapter 2. Figure 6.1 performs this task visually, locating the allocations that have been discussed within the existing framework. As the most intrinsic allocation, grazing permits are located in the lower left quadrant. They were allocated based on a less redistributive and more prepolitical set of entitlements (rewarding individual labor) than was found in the acid rain or climate change cases. More than 50 years later, Congress allocated sulfur dioxide allowances by balancing the same kinds of arguments but generally favoring instead the instrumental notions of greater redistribution for the general good. As amalgamations of intrinsic ideas with substantial instrumental adjustments, both allocations represent a fractious holism that places them within the shaded Hegelian area toward the center of the framework.

The location of greenhouse gas emissions permits is more speculative given that the allocation process remains incomplete. The primary conflict

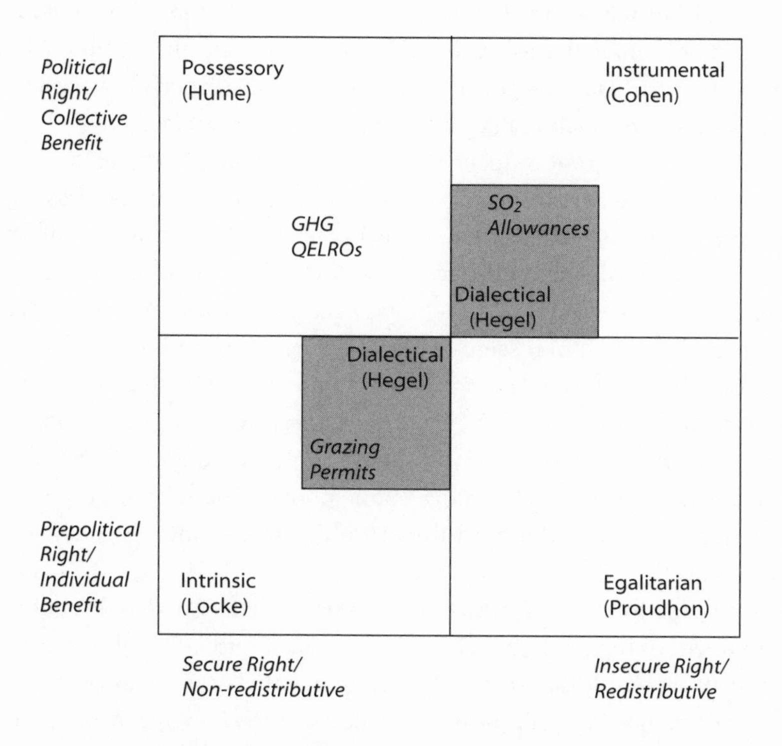

Figure 6.1 Location of Cases within Property Theory Framework

is clearly along the alternative axis, however, primarily between possessory and egalitarian principles. Furthermore, the outcome of the process leading to the Kyoto Protocol favored the Humean position, and thus, allocations of emissions reductions for developed nations are appropriately located in the Humean quadrant. What remains uncertain is where the final allocation of such rights will end up, and whether any other property theory will assume a Hegelian role in synthesizing the two opposing views. If and when such an outcome occurs, another set of shaded boxes would have to be added to the other two quadrants of the framework to encompass this parallel but distinctive outcome.

IMPLICATIONS AND FUTURE DIRECTIONS

At the most basic level, the purpose of this book has been to convince the reader of a deceptively simple idea: equity matters. The assertion is not that equity ought to matter, for few would contest that, or that ideas of equity affect certain personal choices in life that lie outside the public realm. Instead, the conclusion affirms the seemingly idealistic perspective offered in the book's introduction: that as an empirical matter of fact, equity-based norms have a significant influence over environmental policymaking.

The intervening pages have pursued this argument in a variety of settings and drawn a number of additional conclusions along the way, but always in the hopes of convincing the reader of this basic thesis. For a work on public policy, however, the argument has been based on some very abstract and ethereal discussion. It seems worth pausing at this point, therefore, to reflect on what the reader is to make of all this talk of Hegel, property, and equity in environmental policy. Even if we are convinced of the relevance of the property theory framework to these cases, one might ask, what are we to conclude in terms of making public policy? And where, in the end, might the policy scholar or practitioner with an interest in equity concerns go from here?

The answers to many of these questions have been sprinkled throughout the argument of the book. But given the abstract nature of the discussion at times, they are worth isolating and reiterating at this point in a concise format. In short, there are four primary implications that emerge from the successful application of the property theory framework to these cases.

Licensed Property Is a Common and Useful Environmental Policy Tool

One thing this study does *not* conclude is that market-based policies are misguided or inappropriate tools for environmental problems. There is no ethical argument offered here against creating a "right to pollute" or selling "environmental indulgences" (Goodin 1994). Instead, this book recommends a new term—licensed property—to facilitate the use and understanding of market-based policies that create private rights in public resources. Indeed, one reason to recommend the term licensed property is to eliminate some of the apparent squeamishness in certain quarters about owning portions of nature. This is not an expression of the free-market environmentalist argument that the best way to protect nature in all cases is to own it (Anderson and Leal 2001). Recognizing that there are valid arguments against owning some parts of the natural world, in the end this work does not seek to engage the controversy over when owning nature is appropriate or inappropriate. Instead, it argues that when we *do* decide to create private rights in public resources, as we sometimes should, we should be more forthcoming in calling those rights what they are: a form of property, albeit a property with less security than traditional versions found in American jurisprudence.

The use of a clear term like licensed property has several advantages in describing the kinds of private rights in public resources created in the grazing, acid rain, and climate change cases. A single term analytically unifies policies that create grazing permits and emissions allowances, not to mention individual transferable quotas for fisheries and other similar arrangements. All are policy instruments that create rights much like private property. Yet scholars and policymakers working on these issues have only rarely noted the connections between them. The use of *licensed property* to describe the instrument created in all of these cases makes the link between them more explicit and thereby facilitates analyzing and designing new policies by learning from past experience.

The idea of licensed property also refocuses the debate over the type of right being created. Currently, that debate tends to collapse into a stark yes-or-no question: Is it property? This binary approach polarizes the discussion: in answering the question, policymakers are often driven to straddle the issue with their rhetoric, extolling the security of the new right in one sentence ("Yes, it is property") while trying to limit it dramatically in the next ("No, it isn't"). The result is confusion and uncertainty among lawmakers and resource users alike.

In fact, the policy in each of these cases was explicitly designed to capture many of the features of private property to meet specific equity, efficiency, and environmental goals. The resulting property rights may have been weakened in certain ways, but they remained a recognizable form of private ownership. The idea of licensed property would facilitate discussion about the type of right being created by dropping the yes-or-no question in favor of a more sophisticated one: what type of property is this? Such clarity might reduce confusion among policymakers and would certainly provide less room for overheated rhetoric by advocates on both sides. It would shift the discussion from the nature of such rights to the merits of granting or withholding various specific powers of ownership for resource users.

A grazing example illustrates how focusing on the yes-or-no question tends to obscure the details of the conflict and lead to entrenched and irreconcilable positions on either side. Modern critics of the security of land tenure enjoyed by public range users can be both vociferous and simplistic when they respond to the question, "Is a grazing permit property?" by simply saying no. Similarly, advocates for ranchers are sometimes equally simplistic when answering in the affirmative. The result is a policy impasse, with the Bureau of Land Management administrators caught in the middle. The answer to the question may be "no" with respect to the narrow question of takings law, however, the larger policy-relevant reality is much more complex. Grazing permits have always been intended to embrace many of the powers of private ownership without adopting all of them. That point is lost, however, by ending discussion with monosyllabic answers. The idea of a licensed property right would preempt oversimplification and push the conversation in a different direction by more accurately describing the policy originally crafted by legislators and administrators.

Finally, the idea of licensed property could help academics and policymakers better understand and discuss the equity implications of their actions. By explicitly calling these policy instruments a form of property, the term encourages the use of property theory to describe the ensuing allocation arguments and rules. Property theory provides a subset of ideas about distributive justice that is especially relevant to these policy settings. Nor does the set of relevant ideas have to be exhaustive; in the cases studied here, the ideas of just a few prominent property theorists effectively describe and categorize most arguments made by resource users and policymakers alike. Thus, property theory is an excellent tool to begin understanding how equity ideas inform market-based policies in practice.

Allocation Options for Licensed Property May Be Limited

Many discussions of market-based policies note that the initial distribution of rights is independent of the choice to use a property-based instrument in the first place. The pattern followed in all three cases considered here, however, indicates that these two issues may not be independent after all. If certain allocation arguments are as influential as this book suggests, the choice of a licensed property instrument to address an ecological problem *may imply a specific pattern of resource distribution*. Specifically, allocation rules that are inconsistent with certain dominant norms of property allocation have far less chance of adoption.

For example, the Hegelian nature of the allocations in both U.S. examples is striking, given the large contextual differences between the two cases. The acid rain and the grazing allocations varied in terms of the historical period, the existing patterns of prior use, the branch of government primarily determining the rules, and the type of resource being allocated. Yet the process and outcome in both cases were similar to the Hegelian ideal described in Chapter 2, suggesting that Hegelian norms may be quite influential in other U.S. allocations as well. Other contextual factors—especially the degree and quality of prior resource use—further limit the range of plausible allocation options. Those advocating a market-based policy using licensed property, therefore, may have to accept that such a choice implies a relatively specific allocation strategy, given the existing norms in that sociopolitical setting.

This is an important finding for future resource management problems. Previously, supporters of a property-based approach could argue that any distribution of rights under such a policy instrument remained possible—a powerful argument against those protesting licensed property policies on equity grounds. But if, realistically, the range of plausible allocations is much narrower, equity-based opposition to the licensed property approach gains credibility. Those who condemn a distribution with significant intrinsic entitlements, for instance, may have a very good reason for rejecting a property-based approach for resources with substantial histories of prior use, particularly in the U.S. context.

Arguments in favor of auctioning licensed property rights, for example, can expect continued resistance because of powerful intrinsic property norms. Auctions have been a staple of the economic and public policy literature on market-based policies for years, yet few environmental policies have embraced the idea. As a strongly instrumental approach to allocation, auctions run afoul of opposing Lockean principles of entitlement. They also favor the wealthy, who have a correspondingly greater ability to pay for any licensed property being sold. Thus they raise serious equity-based objections in situations with strong histories of "beneficial" prior use (like the grazing case) or substantial inequalities of wealth among potential bidders (like the climate change case). Given the apparent influence of these equity norms, auctions may continue to have a hard time gaining political acceptance in similar policy settings.

On the other hand, allocations based on grandfathering to current users are far less politically influential than they have been portrayed. In fact, the grandfathering of rights based on the status quo is neither common nor inevitable in market-based policies. Although frequently used to describe allocations of licensed property, the term fails to capture fully the actual allocation process in any case considered here. In fact, pragmatic arguments in favor of a distribution based on current possession of the resource are remarkably absent from two of the three cases and tell only part of the story in the third. Both the acid rain and the grazing cases mandated extensive changes from status quo patterns of use, and arguments in favor of a substantial global redistribution of greenhouse gases remain a prominent policy option. Those seeking allocation alternatives to grandfathering should find this outcome heartening, since one can get the impression that allocations deviating from the status quo are politically impossible. In fact, significant adjustments from current use patterns are common, particularly in cases where the type of use is not widely seen as beneficial.

Thus, the range of plausible allocations for a licensed property policy may be both larger *and* smaller than it has been portrayed to date. Viable options clearly exist beyond a basic, default grandfathering to resource users based on current use levels. Yet those viable alternatives must still conform to a relatively limited set of property-based norms, based in large part on the specifics of the allocation context. It appears that there will be little chance of any radical redistribution of use rights in a U.S. case with a lengthy history of beneficial prior use, for example, and few successful arguments based on Locke in a case where prior use is widely judged to be harmful or is absent altogether.

Equity Norms Shape Political Behavior

Chapter 1 reviewed at some length how current theories of political behavior grounded in the rational choice tradition have had difficulty fully understanding various distributive political outcomes. When faced with empirical examples of government action, rational choice theories have failed to provide a fully satisfactory explanation of the observed results. Indeed, several leading scholars in the field have openly pointed to norms and ideology as promising additional explanatory factors.

The results of this work clearly support the conclusion that norms influence policy outcomes and political behavior. Indeed, the primary lesson has been that equity norms have significant influence over the creation and implementation of public policies. In this regard, this book hopes to have made a modest contribution toward the creation of more effective models of political behavior that go beyond the "thin" versions of rational choice theory. In fact, "thicker" models of political behavior considering a wider range of factors are required, and this work has tried to add a little girth our understanding of the policy process in just this manner.

A more elusive goal in understanding political behavior, however, is describing the content of particular norms that apply to specific policy choices. The property theory framework used in this study is an attempt to provide one such description, specifying some of the norms that are influential in at least one particular political context: the creation of market-based environmental policies. Not only do norms matter, in other words, but also in certain situations we can hypothesize and predict which norms are likely to be more or less influential in shaping the political outcome.

In particular, equity norms in these cases prominently affected the initial allocation of the new, licensed property rights to use the resource in question. Therefore, another implication regarding norms and political behavior is that the political process of initial allocation merits closer scrutiny. Although allocations may be less important in terms of the efficiency and ecological goals of such policies, they are crucial issues in the minds of policymakers and resource users. It is important to remember that both the Clean Air Act Amendments and the Taylor Grazing Act nearly failed to become law because of equity-based disagreements, and that climate change negotiations remain threatened by the same kinds of conflicts. Normative ideas appear to be critical in shaping and resolving these disputes. More empirical attention to how allocation processes have worked in other settings therefore seems an important research priority.

Similarly, better clarity and understanding among scholars, if not policymakers themselves, regarding the specific property norms used to justify entitlements to these new rights would be helpful. At present, most allocation arguments in the United States revolve around either intrinsic or instrumental notions of ownership. Yet explicit mention of these theories is entirely missing, even in academic treatments of the subject. Instead, scholars studying market-based policy tend to disregard the property-theory foundations of the allocation question. More empirical research unearthing these normative underpinnings would be a welcome addition. At present, advocates for a particular allocation rule often present their arguments in black and white terms, as if any principle other than their own would be absurd. In reality, extensive ethical and legal justification exists for many such principles, and more scholarship that considers and applies these ideas in a policy context is sorely needed.

In fact, more attention to property theories that attempt an integration or synthesis of conflicting viewpoints seems especially worthwhile—both theoretical explorations of the property theories of writers like Hegel and Kant, as well as more empirical investigations of how their ideas apply in the real world. Given the apparent pattern of conflict between polarized views in allocation debates, more scholarship on property norms and theories that embrace and partially resolve such conflict should be a priority.

Other Licensed Property Examples Merit Further Study

The three cases discussed in this work only suggest the full range of licensed property allocations worthy of study. As reviewed in Chapter 1, governments frequently create and allocate private rights in public resources. Many of these cases would make outstanding additional applications of the property theory framework developed here, both to test its relevance in other contexts and to refine our understanding of where and why particular property norms are most influential. In addition, the framework itself may bring new insights to specific allocation cases that have not yet been considered from this theoretical perspective.

The framework helped spotlight, for instance, how the initial process of administering grazing licenses and permits under the Taylor Grazing Act was heavily influenced by property norms and concerns over fairness. Analysts of this period have tended to overlook this quality, concluding instead that the Division of Grazing was simply the pawn of powerful cattle barons. In reality, this story of "agency capture" by private interests is incomplete at best and fundamentally misleading at worst for the 1934–1938 period. While dedicated to the idea of local control, the Division of Grazing struggled to reconcile the demands of large and small range users based on a wide variety of conflicting entitlements to the public domain. The results recognized the validity of many—but by no means all—of these arguments in reaching a careful amalgamation of conflicting principles and positions. Only when we apply the property theory framework does this significant new perspective on the grazing case emerge. Other allocations may have similarly overlooked qualities and would benefit from examination through this particular theoretical lens.

International examples from the global commons are equally promising candidates for applying the property framework. Chapter 5 already demonstrated how the property norms approach helps explain and interpret the ongoing allocation process in the climate change case. But the potential utility of the framework is hardly limited to this one international process. Other portions of the global commons—the Antarctic continent, ocean bed minerals, high-seas fisheries, objects in outer space—are all potentially subject to future market-based policies and allocation disputes. Applying the framework to future allocations along these lines could be equally illuminating.

For example, many resources in the global commons remain virtually unexploited, lacking any significant prior use. Given the importance of prior use in the allocation process (tangible or intangible, beneficial or not), one would expect this pristine resource condition to imply substantially different allocation principles. In fact, this is already the case: portions of the global commons have been declared by treaty to be the "common heritage of mankind" (CHM). First suggested in 1967 by Arvid Pardo, the United Nations ambassador from Malta, the CHM idea remains ambiguous in international law even after 35 years of use (Baslar 1998). Essentially, however, it stands for a rejection of any private rights to the resource in question, supporting instead some (possibly unspecified) form of resource protection and development with benefits distributed to all nations regardless of priority, location, or other factors (CRS 2000). The CHM idea thereby rejects a property-based approach, or any other private appropriation, of the resources in question (Taylor 1998).

The CHM idea adds new complexity to the question of allocating the global commons. One might conclude that policymakers in future resource allocations will face an additional option of rejecting all private rights in favor of some form of communal resource control. So far, however, CHM status has been proposed only for a few resources lacking any prior use at all. Indeed, it is hard to imagine such an idea taking hold in cases with extensive prior use patterns, given the apparent influence of Humean and Lockean norms.

This apparently unique condition for a CHM argument forces one to ask, what will happen when development pressure on the global commons increases? As the economic value of deep-sea minerals or other resources goes up, for example, or as more affordable technologies for developing these resources are invented, one might well expect stronger challenges to the CHM ideal by private companies and sovereign nations. Two outcomes then seem possible. One is that property-based norms could gain authority as different nations reject the CHM idea and allow private actors to begin developing these resources. This would then put these resources into a situation comparable to those already discussed and make the various principles in the property theory framework more relevant. Alternatively, the CHM idea could serve as a more durable bulwark against unrestricted private development. Treaties incorporating the concept could hold steady or even gain strength, preventing unilateral assertions of ownership or national sovereignty and holding off property rights arguments in general. The potential conflict makes these cases well worth watching.

CONCLUSION

In the end, readers of this work are urged to return to its most fundamental point: equity norms exert substantial influence over the making of environmental policy. Indeed, it seems hard to imagine how one could explain the allocation processes described in this book without recourse to normative principles grounded in property theory. Norms influence our daily lives in myriad ways; why would we expect them, in the end, to be any less influential in our political institutions?

Despite the apparent importance of these normative ideas, however, there is still some reluctance to consider equity explicitly in studies of public policy. Although the trend may be experiencing the beginnings of a welcome reversal, unfortunately it is still the case that for many public policy scholars, equity remains a concept of only secondary importance. Yet, as economist H. Peyton Young (1994, 1) has observed,

> Set against these arguments [against the relevance of equity] is the fact that people who are not acquainted with them insist on using the term "equity" as if it did mean something. In everyday conversation we discuss with seeming abandon the equities and inequities of the tax structure, the health care system, the military draft, the price of telephone service, and how offices are allocated at work. For a term that does not exist this is a pretty good beginning. One is tempted to say that rumors of equity's nonexistence may have been somewhat exaggerated.

One might add the allocation of rights in market-based environmental policies as another example in which equity is discussed extensively despite its limited role in public policy and political science scholarship. Although equity ideas are potentially difficult to discuss, their central role in many policy questions is hard to dispute. Given this fact, the need for even greater attention to the role of equity in public policy remains clear. Politicians and their constituents talk about equity all the time. Why, one might ask, should we who aspire to understand and advise them be reluctant to do the same?

NOTES

Chapter 1: The Politics of Licensed Property

1. This is not to imply that taxes and tradable permit schemes lack any important differences. Taxes fix the total cost to society of the environmental regulation, and then allow the degree of environmental improvement to vary. Permits reverse the priorities. Thus, policy-makers seeking a fixed cost for their policy should turn to taxes, whereas those seeking a fixed level of environmental improvement might favor permits. In addition, a permit approach that includes trading can further improve efficiency when there is a relatively large variation among users in the marginal cost of complying with the regulations. On this last point, see Stavins 1997, 16.
2. Many fisheries are open for harvest on a certain dates and remain open only until the total allowable catch is achieved. This encourages a race in which those who catch fish the fastest get the largest share. Fishermen make inefficient investments in large boats and equipment and engage in dangerous practices of fishing around the clock. As a result, an entire year's catch is harvested in a very short time, often just a few days (Grafton 1996, 7).
3. This is especially true for individual transferable quotas. For a striking example of this oversight, see "Real World Results" (a pamphlet published by the Emissions Trading Education Initiative and distributed by the Environmental Defense Fund [now Environmental Defense]), which argues that the "essential change" ending the race for the fish was the conversion of a capped market to a cap-and-trade system. This is somewhat misleading: the essential change was the granting of more secure property rights to a limited group of fishermen. The power to transfer is irrelevant to this point.

4. One exception is the New Zealand individual transferable quota program, which creates a *permanent* right to a specific share of the total allowable catch each year (Annala 1996, 45). This is no longer "licensed" property—it is simply a property right.
5. Chief Justice William Rehnquist, for example, calls the right to exclude others "one of the most essential sticks in the bundle of rights that are commonly characterized as property." See *Dolan* v. *City of Tigard,* 512 U.S. 374 (1994), quoting from *Kaiser Aetna* v. *United States,* 444 U.S. 164, 176 (1979).
6. This is confirmed by the reluctance of many advocates of market-based policies to support "open market trading" programs lacking a secure cap on the total number of rights granted. See written testimony of Dan Dudek, Hearing before the Joint Economic Committee, U.S. Congress, July 9, 1997, at 11. The Environmental Protection Agency (EPA) continues to permit such rules, however, under Title I of the 1990 Clean Air Act Amendments; see U.S. EPA, *Draft Economic Incentive Program Guidance,* September 1999, at 16–18.
7. This has been observed by employees of the General Accounting Office, among others. See Air Pollution: Overview and Issues on Emissions Allowance Trading Program (Testimony before Joint Economic Committee, U.S. Congress, July 7, 1997), GAO Report T-RCED-97-183 (1997), at 11.
8. A similar but now extinct form of licensed property was the unpatented homestead on public lands.
9. Holders have to pay $100 per year to the government or otherwise "develop" their claims annually, for example, to keep the claims from lapsing.
10. Interestingly, as an exception to this rule, the New Zealand ITQ program began with stronger property rights than the Clean Air Act Amendments created. Each ITQ holder was entitled to a specific tonnage of fish rather than a percentage of some floating total. To reduce fishing pressure, the New Zealand government thus had to buy back ITQs from the holders. This plan was adjusted to a percentage of the total allowable catch after several years of government buybacks that were no longer deemed politically possible. It is also interesting to note that compensation for reductions in the total catch were paid to ITQ holders for five years after the transition, as part of the negotiated settlement to move to a percentage basis (Annala 1996, 55–56).
11. One clear candidate is the right to an airplane gate at major national and international airports (Keohane et al. 1998, 353; *The Economist,* August 22, 1998,348). For a more unorthodox example, consider the allocation of rights to raft on specific federal rivers, such as the Colorado (*High Country News,* December 21, 1998, 1).
12. The now-famous Coase Theorem demonstrates that any clear allocation of property rights would lead to an efficient distribution of resources, under specific market conditions. The implication is that clear and complete property rights will reduce social costs, regardless of how they are assigned.
13. Where *efficiency* is defined in general terms as attaining the greatest aggregate social value in society; see discussion at end of this chapter.
14. More formally, an externality of this type means the marginal cost curve of the firm is too low: it does not reflect the full social cost of producing a given amount of product.
15. See Michael J. Sandel, It's Immoral to Buy the Right to Pollute, *New York Times,* A-23 (December 15, 1997); Peter Passell, Economic Scene: Selling Pollution Rights Isn't Popular; Neither Are Alternatives, in *New York Times,* April 8, 1993, D2; or Editorial, The Wrong of Pollution Rights, *Boston Globe,* September 19, 1989. See also Rosenberg (1994, 525–26), Goodin (1994), and Sagoff (1988).

16. In 1997, for example, all affected utilities were in compliance with the emissions rules, and total emissions of SO_2 for the year were 1.7 million tons (23%) below the cap of 7.1 million tons. U.S. EPA, *1997 Compliance Report: Acid Rain Program*, EPA-430-R-98-012 (1998).
17. The question of fees for public lands grazing permits has been controversial since federal regulation of the range began in 1897. Although many studies argue for or against a given scale of fees in terms of fairness, few attempt to define *fair* with any precision. The notion of a fair fee, of course, depends on the equity principles that guided the allocation of the range in the first place. In this respect, the current study might provide some new insight into a very old controversy by tackling a question prerequisite to the fee debate: on what basis did the government assign the rights for which fees are to be charged?

Chapter 2: A Property Theory Framework

1. A portion of Sections I and III of this chapter previously appeared in a substantially different form in Raymond and Fairfax (1999).
2. Most theories of property focus primarily on the notion of private property. Given this fact, and the focus of this book on private, licensed property rights, in this chapter the terms *property* and *private property* will generally be used interchangeably. Where *property* specifically does not refer to private property, the distinction will be made clear.
3. This is well discussed by Horwitz (1992), and others. For a summary on this point, see Raymond and Fairfax (1999, sec. III).
4. Waldron (1994, 115) notes with concern that if we accept the Humean argument as a moral justification for private property, we are stuck with the initial distribution of goods regardless of how imbalanced it might be. This is a weak moral vision of equity at best: "Why," asks Waldron, "should we expect heavily moralized standards like justice and fairness—standards that connote the idea of the rightfulness of the proportion of one person's holding to another's—to emerge from the essentially amoral process that Hume ... describe[s]?"
5. Why this is so obvious to Locke is unclear to some later commentators (e.g., Becker 1977, 36–41).
6. Locke notes in this section, therefore, the impossibility of taking ownership of land by unilateral action in England, where "enough and as good" certainly no longer existed for others.
7. The use of the term *instrumental* to describe a theory of property along these lines is well established in the literature. See Horwitz (1992) for a leading example. This work uses the term in the same general spirit as these previous works, but some of the details will vary.
8. The Progressive scholar Wesley Hohfeld (1919) noted that rights and duties can be viewed as "legal correlatives" of one another: the notion of a right for one person necessarily implies a duty in others to the right holder to forbear or take certain action. The notion of rights and duties as legal correlatives raises a host of possible objections and complications, including the notion of rights without corresponding duties and others. For a brief introduction to these complications, see Waldron (1988, 68–73).
9. Rousseau's theory of property would be another prominent example, as would the modern work of environmental legal thinker Joseph Sax.
10. Cohen wrote widely on a variety of topics during his career, including law, philosophy, and science. Several good collections and works include *Reason and Nature* (1959), *Studies in Philosophy and Science* (1949), *The Faith of a Liberal* (1946), and *Law and the Social Order*

(1967). His most famous essay, "Property and Sovereignty," is cited here as reprinted in the *Law and the Social Order* collection.

11. This principle of Hegel's account has profound implications for policy related to the natural world that merit much more consideration than can be given here.
12. All property is "alienable" in Hegel's account by virtue of an individual's freely choosing to release his ownership claim (sec. 65).
13. Peter Stillman (1980) discusses the importance of property to individual development within Hegel's work more generally.
14. The parallels between this observation by Hegel and those of modern legal scholar Carol Rose are quite thought provoking. In Rose's book of essays *Property and Persuasion*, she concludes by arguing that in many important ways, the key to understanding a property claim is to think of it as an act of persuasion to the outside world (Rose 1994, 297). The connections to the ideas of Hegel are clear.

Chapter 3: Allocating SO_2 Emissions Allowances, 1989–1990

1. U.S. Senate Committee on Environment and Public Works, *Report on S. 1630* (December 20, 1989), at 289 (hereafter cited as SCEPW Report on S. 1630).
2. One Btu is the amount of energy required to raise the temperature of one pound of water by one degree Fahrenheit.
3. I am greatly indebted to the guidance of Gary Hart of Southern Co. on the complexities of Btus consumed and MW of electricity generated.
4. The literature on the possible ecological implications of acid rain is extensive. Two excellent general articles on the ecological aspects of the problem are Mohnen (1988) and Bricker and Rice (1993). The federal government also sponsored an intensive, 10-year study of the problem during the 1980s called the National Acid Precipitation Assessment Program (NAPAP). The *1990 Integrated Assessment Report* (NAPAP 1991), though scientifically exhaustive, was largely irrelevant to the design and implementation of the Title IV program for a variety of political reasons.
5. The construction of tall stacks to meet local air quality standards is also alleged to have been illegal under the law despite having been approved by EPA; see SCEPW Report on S. 1630, at 289.
6. Kete (1992a,134) notes that in 1986, more money was spent lobbying Congress on acid rain than on any other issue.
7. Environmentalists were opposed because they believed—correctly, it turns out—that they could get a better deal after the 1988 elections. Industry thought—incorrectly—that it could continue its successful battle to delay any regulation on the issue at all.
8. EDF internal memo dated March 24, 1989.
9. EDF internal memo dated March 23, 1989.
10. EDF internal memo dated March 23, 1989.
11. In the exact language of the Bush proposal, "An allowance issued under this title is a limited authorization to emit sulfur dioxide or nitrogen oxides in accordance with the provisions of this Act. Such allowances may be limited, revoked, or otherwise modified in accordance with the provisions of this Act, or future amendments to it." House Resolution 3030, Section 503(f), at 238.

12. As stated, for example, in Bush's letter to Senate Minority Leader Robert Dole, dated January 19, 1990, as printed in the Senate Floor Debate 1/90, reproduced in *A Legislative History of the Clean Air Act Amendments of 1990* (hereafter cited as LHCAAA), at 4,935.
13. House Resolution 3030, Section 504, at 238–48.
14. House Resolution 3030, Section 504, at 233.
15. This method is not specifically indicated in the law but is the basis for the calculation of phase I allowances in Table A under the act. See House Resolution 3030, Section 504, at 241–47.
16. For example, a plant that burned 100,000 million Btus of energy in 1985 and emitted at a rate of 5.0 pounds per million Btus would be entitled to 500,000 pounds of SO_2 emissions, or 250 allowances, under the EDF plan but only 250,000 pounds, or 125 allowances, under the Bush modifications. The math: total emissions entitlement = million Btus consumed ¥ actual emission rate under EDF proposal = 100,000 million Btus ¥ 5.0 pounds per million Btus = 500,000 pounds, or 250 tons, of SO_2 emissions. Under the Bush plan, the rate drops to 2.5 pounds per million Btus regardless of actual emissions rate, cutting this hypothetical plant's allocation of allowances in half.
17. House Resolution 3030, Section 505, at 247–48.
18. House Resolution 3030, Section 505, at 247–48.
19. House Resolution 3030, Section 511, at 262–64.
20. House Resolution 3030, Section 510, at 261–62.
21. Hearings were also held in the spring and summer 1989 by California Representative Henry Waxman's Subcommittee on Health and Environmental Protection, but they dealt only briefly with the acid rain issue.
22. House Energy and Power Subcommittee Hearing (hereafter cited as HEPSH), September 7, 1989, at 1.
23. Senate Environmental Protection Subcommittee Hearing (hereafter cited as SEPSH), September 26, 1989, at 1.
24. SEPSH September 26, 1989, at 19; HEPSH September 12, 1989, at 7.
25. SEPSH October 3, 1989, at 9. EPA staff affirmed this position in testimony as well; for example, see written statement of William Rosenberg, EPA, Senate Energy and Natural Resources Committee Hearing (hereafter cited as SENRCH), January 24, 1990, at 70.
26. This is not to imply that EDF argued allowances were protected by the Fifth Amendment; only that their descriptions of the nonproperty status of allowances was much milder than those of other environmental groups. See, for example, oral testimony of EDF economist Daniel Dudek, SEPSH, October 4, 1989, at 203.
27. Written testimony, SEPSH, October 3, 1989, at 87.
28. SEPSH, October 3, 1989, at 9.
29. For example, see oral testimony of Ruth Gonze, American Public Power Association, HEPSH, September 7, 1989, at 673 and 698; or written testimony of the same group, SEPSH, October 4, 1989, at 483; written testimony of David Penn, Wisconsin Public Power, HEPSH, October 11, 1989, at 391; or oral testimony of Representative Wise of Nevada, HEPSH, October 18, 1989, at 12.
30. Written statement of James Rogers, Public Service Company of Indiana, SEPSH, October 4, 1989, at 525. See also oral testimony of William Badger, NARUC, SENRCH, January 24, 1990, at 181 and 216.
31. HEPSH, October 11, 1989, at 345 and 350.

32. For one example of the savings-and-loan argument, see written statement of James M. Friedman (for the Coalition for Environmental Energy Balance), SEPSH, October 4, 1989, at 443. For a strong assertion of acid rain as a national problem more generally, see, for example, testimony of Senator John Glenn, SEPSH, October 3, 1989, at 22.
33. HEPSH, September 7, 1989, at 457.
34. HEPSH, September 7, 1989, at 1.
35. Paul Simon of Illinois, for example, asserted the electricity fee would be the "fairest approach" to the cost-sharing issue. See SEPSH, October 4, 1989, at 190.
36. Many opponents insisted that a new tax was blatantly unfair in making clean states pay twice for SO_2 abatement. Members from the West and Southeast were especially vociferous in this argument; see, for example, statement of Representative Bilirakis of Florida, HEPSH September 7, 1989, at 522.
37. For one cogent example of this argument, see Department of Energy official Linda Stuntz's oral testimony, SENRCH, January 24, 1990, at 53.
38. The view was also apparently shared by many utility regulators. See testimony of William Badger, speaking on behalf of the nation's state utility regulators, who noted that although the administration proposal was being touted as a polluter-pays approach with no regional subsidies, many members of his organization believed it represented a form of *de facto* cost-sharing via the higher emissions rates for older, high-emitting plants. SENRCH, January 24, 1990, at 182.
39. HEPSH, October 18, 1989, at 78.
40. Even Representative Cooper, an opponent of a national tax for cost-sharing, was opposed to an egalitarian emissions rate standard. See HEPSH, October 11, 1989, at 344.
41. For example, the Alliance for Acid Rain Control, a group of current and former governors from various states, including those in the West, was tolerant of the allocation of allowances in favor of the Midwest. See written testimony of Governor Roy Romer of Colorado, SEPSH, October 4, 1989, at 393.
42. SEPSH September 26, 1989, at 71.
43. Written testimony of A. Joseph Dowd, American Electric Power, SENRCH, January 24, 1990, at 138.
44. The reasons ranged from economic recession to introduction of new nuclear capacity in 1985 (testimony of Senator Dan Coats, Indiana, SEPSH, October 3, 1989, at 5; and written testimony of William Cornelius, Union Electric Co., SEPSH, October 4, 1989, at 362).
45. Written testimony of William Rosenberg, EPA, HEPSH, October 11, 1989, at 246: "We believe that the 'allocations' should be based on the most recent and accurate data available. This dictates the use of 1985 data for the emission rates and 1985–87 for fuel consumption."
46. Testimony of William Rosenberg, EPA, SENRCH, January 24, 1990, at 226 and 399.
47. HEPSH, September 12, 1989, at 95.
48. HEPSH, September 12, 1989, at 101.
49. HEPSH, September 7, 1989, at 462. The exclusion of industrial sources also had regional fairness implications, since some states with relatively high SO_2 emissions from industry, such as Texas, were left largely unaffected by the acid rain bill's focus on utility sources.
50. Oral testimony of Jon Prendergast, Council of Industrial Boiler Owners, SEPSH, October 4, 1989, at 232.

51. Oral testimony of William Rosenberg, EPA, HEPSH, October 11, 1989, at 353. It is worth noting that this argument makes little sense when applied to industrial boilers used to generate power for manufacturing.
52. HEPSH, October 18, 1989, at 84.
53. Written testimony of Representative Virginia Smith from Nebraska, HEPSH, October 4, 1989, at 189.
54. HEPSH September 12, 1989, at 128.
55. For example, see Administrator Reilly's exchange with Representative Bilirakis, HEPSH, September 12, 1989, at 126-129; or with Representative Neilson, HEPSH, September 12, 1989, at 99–101.
56. Oral and written testimony of Richard Ayres of the National Clean Air Coalition, SENRCH, January 25, 1990, at 375–77 and 381–82.
57. Testimony of Paul Schmechel, SEPSH, October 4, 1989, at 174. Plants burning natural gas were particularly concerned because they relied on the option of burning oil, with higher SO_2 emissions, in case of gas supply interruptions and as a competitive pressure to keep the price of natural gas down. Having burned only natural gas in 1985, however, many such plants had an actual emissions rate from that year so low as to prevent any use of oil in the future. See oral statements by Representative Tauzin, Louisiana, HEPSH, September 12, 1989, at 117; written statement of American Public Power Association, SEPSH, October 4, 1989, at 487.
58. For example, see written testimony of American Petroleum Institute, HEPSH, September 7, 1989, at 709; oral testimony of Harry Storey, Alliance for Clean Energy, SEPSH, October 3, 1989, at 29.
59. Oral testimony of Dan Berube, Montana Power, SEPSH, October 9, 1989, at 566.
60. For example, see HEPSH, October 18, 1989, at 71.
61. See oral testimony of William Badger, NARUC, SENRCH, January 24, 1990, at 181; statement of New York Representative Norman Lent, HEPSH, September 7, 1989, at 529.
62. For example, see oral testimony of Richard Schmalensee, Council of Economic Advisors, HEPSH, October 11, 1989, at 228; written testimony of Daniel Dudek, EDF, SENRCH, January 24, 1990, at 112.
63. Written testimony of Western Energy Supply and Transmission Associates, SEPSH, October 3, 1989, at 544.
64. Oral testimony of Roy Alper, President, Independent Power Corp., SEPSH, October 3, 1989, at 43–45; written testimony of James Kelly, National Independent Energy Producers, HEPSH, October 11, 1989, at 357–70.
65. Kelly, HEPSH, October 11, 1989, at 371.
66. The examples are innumerable. For two good examples, see House floor debate on the bill, May 1990, LHCAAA, at 2,620, or Senate conference committee bill debate, October 26, 1990, LHCAAA, at 1,061.
67. 104 *Stat* 2584, especially at Sections 404–406, which combine to make up 21 pages of the final law.
68. One staff person closely related to the acid rain title of the bill recalled that numerous members of Congress dropped by his tiny office in spring 1990 to make "lofty equity arguments" regarding the allocation rules in the bill. By his estimate, of the 8.9 million annual SO_2 allowances distributed by the bill, more than 8 million were allocated based on rules grounded in equity considerations. Interview with Joe Goffman, EDF, October 1, 1999.

69. Statement of Senator Pete Domenici, Senate debate October 26–27, 1990, LHCAAA, at 1,111.

70. Interviews with Brian McLean, EPA, October 25, 1999; John McManus, American Electrical Power, January 17, 2000; and Rusty Matthews, former aide to Senator Robert Byrd, December 22, 1999.

71. S. 1630 as passed, Section 403(f), in LHCAAA, at 4598. The same phrase remains in the final version of the law: Public Law 101-549, 104 *Stat* 2584 (1990), at Section 403(f).

72. SCEPW Report on S. 1630 at 321.

73. Senator Stephen Symms of Idaho, for example, made this argument more than once during public debate on the bill. Senate Committee Report, at 472, in LHCAAA at 8812; Senate conference debate, October 26, 1990, LHCAAA, at 754; for the reference to Nobel prize-winning economists, see Senate conference debate, October 26, 1990, LHCAAA, at 755.

74. SCEPW Report on S. 1630, at 321.

75. Senate conference debate, October 27, 1990, in LHCAAA, at 1,034.

76. Extended Remarks on Passage of S. 1630, November 2, 1990, in LHCAAA, at 10,766.

77. The original compliance deadline for Phase I was December 31, 1995. The House moved that deadline to January 1 of the same year. Interview with Dirk Forrister, former staff member for Representative Jim Cooper, December 10, 1999. The language can be found in Section 404(a)(2) of the bill as passed by the Senate 4/3/90, in LHCAAA, at 4,599

78. Section 404(d) of the bill passed by the Senate April 3,1990, in LHCAAA, at 4,604–609.

79. House Committee Report on H.R. 3030, at 370.

80. Such new plants were given allowances based on a standard figure of 65% of capacity. See remarks of Reprsentative Pursell of Michigan, House conference debate, October 26, 1990, in LHCAAA, at 1,281.

81. Senate floor debate, March 1990, in LHCAAA, at 7,071–72.

82. LHCAAA, at 7,070. According to the senator, the permitted increase in Texas emissions would offset more than half the reductions required of his state, Illinois.

83. 104 *Stat* 2584, at Section 410. The law makes no mention of industrial sources except to permit them to opt in to the program voluntarily under Section 410. To date, very few industrial sources have exercised this option; see *EPA 1998 Compliance Report—Acid Rain Division.*

84. Interview with Joseph Goffman, EDF, October 1, 1999; interview with Judy Greenwald, former staff member for Representative Phil Sharp, December 21, 1999; interview with Brian McLean, EPA, October 25, 1999.

85. House debate, in LHCAAA, at 2,562.

86. SCEPW Report on S. 1630, at 309; also discussed in Kete (1992a), at 195–201. The final rules are in 104 *Stat* 2584, at Section 405(d), 405(e), and 405(f).

87. Most of these rules can be found in 104 *Stat* 2584, at Section 405.

88. 104 *Stat* 2584, at Section 403(a).

89. For example, see comments of Senator McClure of Idaho, Senate floor debate 4/90, in LHCAAA at 7,165; or Representative Kolbe of Arizona, House floor debate 5/90, in LHCAAA at 2,622.

90. 104 *Stat* 2584, at Section 405(d). Several members of Congress made explicit reference to the importance of this shift in the final public debate over the bill. See remarks of Representative Bilirakis of Florida, House conference debate, October 26, 1990, in LHCAAA, at 1,248; remarks of Senator Nickles of Oklahoma, Senate conference debate, October 27, 1990, in LHCAAA, at 1,061.

91. SCEPW Report on S. 1630, at 304–305. Note that the purpose of the auction was twofold: to provide access for new entrants to the market and to provide some price signals to help stimulate private trading.
92. Interview with Judy Greenwald, December 21, 1999; interview with John McManus, AEP, January 17, 2000. As one former congressional staff member recalled, the process was focused on trying to solve the problems of current operators, which were deemed more compelling than those of new entrants. Interview with Dirk Forrister, December 10, 1999.
93. 104 *Stat* 2584, at Section 405(g).
94. The goals and priorities for administrators to carry out in other significant environmental statutes have often been unclear. The Federal Land Policy and Management Act (FLPMA), for example, gives its implementing agency (the Bureau of Land Management) very little beyond a general instruction to pursue a "multiple-use mandate"; see Dana and Fairfax (1980). Some commentators have publicly praised the detailed allocation of allowances by Congress; see Environmental Law Institute (1997). The director of the EPA acid rain program is less sanguine about the role of Congress in allocation (McLean 1997).
95. There were 0.9 million allowances involved in economically significant trades according to EPA in 1994, versus 12.7 million allowances traded in 2000. U.S. EPA Acid Rain Program *2000 Annual Progress Report.*
96. Interview with Gary Hart, Southern Co., November 3, 1999.
97. Described in 104 *Stat* 2584 (1990), at Section 404.
98. Interview with Brian McLean, EPA, October 25, 1999; interview with Rusty Matthews, December 22, 1999.
99. Florida, for example, was a net seller rather than a buyer of allowances as of 1999. Interview with Brian McLean, EPA, October 25, 1999.
100. Interview with Gary Hart, the Southern Co., November 3, 1999.
101. EPA Acid Rain Program website, *www.epa.gov/acidrain.* For a news account of one such group, see Davidson Goldin, "Law Students Buy and Hold Pollution Rights," *New York Times,* March 31, 1995, A28.
102. Interview with Gary Hart, Southern Co., November 3, 1999; interview with John McManus, AEP, January 17, 2000.
103. See Maria L. LaGanga, "Emissions Trading Plan Is Slow to Get into the Air," *Los Angeles Times,* May 27, 1993, A1.
104. Matthew L. Wald, "Suit Attacks Swap Plan on Pollution," *New York Times,* March 14, 1993, A35.
105. Statement of Senator James McClure, Idaho, SENRCH, September 7, 1989, at 430. (Senator McClure gives original credit for the remark to Senator Ford.)
106. For example, see statement of William Reilly, SENRCH, September 26, 1989, at 94; statement of Senator John Heinz, Pennsylvania, SENRCH, October 3, 1989, at 7.
107. Senate floor debate, March 1990, in LHCAAA, at 6,781.
108. One or two persons were exceptions, recognizing the existence of previous models for the acid rain program. See Department of Energy representative Linda Stuntz, for example, SENRCH, January 24, 1990, at 224. Even they, however, failed to make connections to several examples mentioned in this research, including the Taylor Grazing Act case that is the subject of Chapter 4.

Chapter 4: Allocating Public Lands Forage, 1934–1938

1. A small portion of the material covered in Sections III and IV of this chapter was previously published in a substantially revised format and context in Raymond (2002), Localism in Environmental Policy: New Insights from an Old Case, *Policy Sciences* 35: 179–201.
2. The terms are not interchangeable and will be distinguished in the discussion ahead.
3. Assistant Secretary of the Interior Theodore Walters, Memo for the Grazing Policy Committee, April 2, 1937 (National Archives, Washington, D.C., Record Group 49, Entry 3, Box 2; hereafter cited as NA-DC, RG-Entry-Box). For another allusion to the ongoing need for fairness in allocation, see also Walters's speech, transcript, Wyoming Wool Growers Meeting, August 5, 1937, at 51 (American Heritage Center, Entry 1350, Box 5; hereafter cited as AHC, Entry-Box).
4. Rowley (1985, 91) notes that Forest Service permittees got around the maximum limit by leasing stock to others who managed them within the reserve. The Forest Service did not dispute this practice even though ownership of the stock remained with the original permittee in excess of the maximum limit.
5. The limit was a 3% reduction per year. Actual reductions in many cases may have been even smaller (West 1982, 47).
6. Rowley (1985) notes that increased homesteading under the Stockraising Homestead Act of 1916 contributed to this situation. Roberts (1963) adds that the number of small local livestock associations went from about 90 in 1911 to more than 600 by 1919.
7. The controversy led to congressional hearings and a bill sponsored by Representative Stanfield of Oregon that would have created nearly full legal property rights in Forest Service permits. Although the Stanfield bill failed to pass, the agency made many of the same changes in its regulations under intense pressure from stockmen and legislators alike. See West (1982).
8. Although the American National Livestock Association (ANLA) passed a number of resolutions calling for some form of leasing of the public domain in the 1920s, sheepmen remained divided on the issue all the way through the consideration of the Taylor Grazing Act in 1934. See Resolution 1, 1927 ANLA annual conference (AHC, 1713–14). For another example of the ongoing discord within the sheep industry, see The Public Domain Bill, in *The National Wool Grower*, April 1932, at 7 (in which it is noted that the National Wool Growers Association has taken no position on the most recent public lands bills because of internal disagreement among their state-level affiliates).
9. Calef (1960). As one Wyoming attorney noted in 1929, "... so far as the public land states are concerned, [the devolution issue] will be the overshadowing political question until it is rightly settled." *Wyoming Stockman-Farmer* 35(12): 1, Cheyenne.
10. Peffer (1951). Privately, it would seem that such conflicts continued unabated. Ickes appeared to take seriously at least one suggestion from his staff in 1934 that transfer of the Forest Service to the Department of the Interior could be a part of the TGA. Memo, Rufus Poole to Secretary Ickes, March 13, 1934, National Archives II, College Park MD, Record Group 48, Entry 812, Box 1 (hereafter cited as NA-MD, RG-Entry-Box).
11. Peffer (1951). The idea for nonlegislative regulation of the public domain by executive action is attributed originally by Martin Mayer to Rufus Poole of the Interior Solicitor's Office. See Mayer, *The Lawyers* (1966, 354–55). I am indebted to Jim Muhn of the Bureau of Land Management for this reference.
12. 48 Stat. 1269.

13. Statement of Representative Burton L French of Idaho, Hearings on H.R. 11816 (the Colton Bill) before the House Committee on the Public Land, June 2, 1932, at 147 (hereafter cited as Colton Bill Hearings).
14. Oral testimony of Secretary of the Interior Harold Ickes, Hearings on H.R. 6462 (the TGA) before the House Committee on the Public Lands, June 7, 1933, at 13 (hereafter cited as House TGA Hearings).
15. Written Testimony of R. Y. Stuart, Forester, in Grazing on the Public Domain, Colton Bill Hearings, May 31, 1932, at 92.
16. Oral testimony of Representative Ed Taylor of Colorado, House TGA Hearings, June 8, 1933, at 34.
17. Comments by Senator King, Hearings on H.R.6462 (the TGA) before the Senate Committee on Public Lands and Surveys, April 27, 1934, at 89 (hereafter cited as Senate TGA Hearings). See also comments of Representative Chavez of New Mexico, House TGA Hearings, June 7, 1933, at 16, 29.
18. Indeed, testimony during the hearings noted the absence of specific guidance on allocation issues within the bill. See letter from W. P. Wing, Secretary of the California Wool Growers Association, House TGA Hearings, March 1, 1934, at 192. One speaker went so far as to say that such an allocation of grazing rights could not "be reasonably done in any bill." See statement of Herman Stabler, U.S. Geological Survey, House TGA Hearings, February 20, 1934, at 87.
19. The stated goals of the TGA include "to stop injury to the public grazing lands by preventing overgrazing" and "to stabilize the livestock industry dependent upon the public range." 48 Stat. 1269, at Introduction.
20. Written testimony, Colton Bill Hearings, May 3, 1932, at 8. Interestingly, virtually the same phrase was used by Ickes in his letter to Congress in favor of the TGA one year later. See House TGA Hearings, June 7, 1933, at 4.
21. Oral testimony, House TGA Hearings, June 8, 1933, at 30–34.
22. See, for example, the verbal exchange between Representatives Mott and Ayers, House TGA Hearings, February 28, 1934, at 168.
23. House TGA Hearings, February 28, 1934, at 168.
24. House TGA Hearings, June 9, 1933, at 43.
25. House TGA Hearings, March 2, 1934, at 196.
26. Oral testimony, Senate TGA Hearings, April 26, 1934, at 55–56.
27. Oral testimony, Senate TGA Hearings, April 27, 1934, at 69–71.
28. 48 Stat. 1269, Sec. 3.
29. The case law was not decided fully until *U.S. v. Fuller,* 409 U.S. 488 (1972). For a summary of the legal and political struggles over the question of property rights in grazing permits, see Raymond (1997).
30. 48 Stat. 1269, at Sec. 3.
31. Representative Taylor was not always entirely specific regarding which rights he was referring to, however. Oral testimony, Senate TGA Hearings, April 26, 1934, at 31; oral testimony, House TGA Hearings, February 19, 1934, at 75.
32. Oral testimony of Northcutt Ely, assistant to the secretary of the Interior, Colton Bill Hearings, May 24, 1932, at 52; for Ickes' statement, see oral testimony, Senate TGA Hearings, April 20, 1934, at 10.

33. Oral testimony, Senate TGA Hearings, April 26, 1934, at 61. Staff from other agencies expressed similar support on normative grounds for recognition of prior use as an allocation principle. See oral testimony of Herman Stabler, U.S. Geological Survey, House TGA Hearings, February 20, 1934, at 88.

34. This was stated most clearly in a letter from W. P. Wing of the California Wool Growers Association: "The Taylor bill does not state to whom permits shall be granted. In other words, no preference is given to present users." House TGA Hearings, March 1, 1934, at 192.

35. Oral testimony, Senate TGA Hearings, April 30, 1934, at 124.

36. Oral testimony of Howard J. Smith, Arizona Cattle Growers Association, Senate TGA Hearings, May 1, 1934, at 171.

37. Oral testimony, Senate TGA Hearings, May 1, 1934, at 184.

38. Oral testimony, Senate TGA Hearings, April 30, 1934, at 150.

39. Arguments at the hearings in favor of equity for small ranchers took an ambivalent position toward the question of prior use, seeing it as both a protection for and a threat against smaller livestock outfits. For prior use as a potential asset to small users, see statement of Frank Bryant, Mesa-Alto Livestock Association of New Mexico, Senate TGA Hearings, May 1, 1934, at 185–86; or statement of J.B. Wilson, Wyoming Wool Growers Association, Senate TGA Hearings, April 30, 1934, at 124–25. For one example of the "threat" view, see the interchange between Dan Hughes, Colorado Wool Growers Association, and Representative Lemke, House TGA Hearings, February 20, 1934, at 96–97.

40. Oral testimony, House TGA Hearings, June 7, 1933, at 10.

41. Oral testimony of Rufus Poole, House TGA Hearings, February 21, 1934, at 128–29.

42. Oral testimony of F. M. Marshall, National Wool Growers Association, Senate TGA Hearings, May 1, 1934, at 189.

43. Oral testimony, House TGA Hearings, February 19, 1934, at 77.

44. 48 Stat. 1269, at Sec. 3. The requirement is quite similar to those written into various homesteading laws.

45. For just one example of the xenophobic overtones in such criticism, see comments of Ed Taylor, Senate TGA Hearings, April 26, 1934, at 28. At least one member of the House expressed concern during hearings that some owners of large flocks were noncitizens. See question during testimony of Dan Hughes, Colorado Wool Growers Association. However, if the rule was directed at excluding Basque and Greek shepherds from the public domain, it may have been ill conceived. Hughes, for example, argued that most of these foreign-born owners were already in the process of becoming naturalized citizens of the United States and, in his estimation, were likely to become "pretty fair citizens." House TGA Hearings, February 20, 1934, at 99.

46. Amended House version of Taylor Grazing Bill, Sec. 3, as reprinted in House TGA Hearings, at 195.

47. 48 Stat. 1269, Sec. 3.

48. 48 Stat. 1269, Sec. 3. The full clause read as follows: "Preference shall be given in the issuance of grazing permits to those within or near a district who are landowners engaged in the livestock business, bona fide occupants or settlers, or owners of water or water rights, as may be necessary to permit the proper use of lands, water, or water rights owned, occupied, or leased by them. ..."

49. Oral testimony of Representative Taylor, Senate TGA Hearings, April 26, 1934, at 28.
50. Oral testimony of Representative Taylor, House TGA Hearings, June 7, 1933, at 14.
51. One other section of the law also favored property owners who were most proximate to the public lands. Section 15 permitted individual leasing of isolated tracts of public grazing land not fit for inclusion in a larger grazing district. The bill permitted the leasing of these tracts only to private property owners adjacent to the public land in question. Prior use was not mentioned. Thus, leasing under Section 15 had an even stronger instrumental, commensurate property requirement in which property owners immediately next to public lands were given an absolute preference over other applicants. See 48 Stat. 1269, Sec. 15
52. See Senate TGA Hearings, April 30, 1934, at 149.
53. Senate TGA Hearings, April 30, 1934, at 149–52.
54. The 80 million-acre limit was later lifted in a 1936 amendment to the bill.
55. "It seems to me," stated Carpenter at the 1935 Nevada meeting, "it is just this: that insofar as the rules of the Secretary of the Interior recognize the customs and usages of that range and adopt them, he is all right, and insofar as it fails to do it, he is all wrong." Transcript, Statewide Meeting for Nevada, January 24, 1935, at 74 (Farrington Carpenter Papers, Western History Section, Denver Public Library; hereafter cited as FRC-DPL).
56. For example, see Taylor Act Administration, *National Wool Grower*, December 1935, at 16; or letter, Milo Deming to C. F. Moore, January 23, 1936, regarding dissatisfaction with actions of the Steamboat Springs Advisory Board (NA-DC, 49-3-13).
57. Carpenter did have some allies in the Washington office, including most significantly Assistant Secretary of Interior Theodore Walters, another westerner.
58. Transcript, TGA Hearing in Cody, Wyoming, August 2, 1934, at 9 (National Archives, Denver, Record Group 49, Entry 45, Box 171; hereafter cited as NA-DNV, RG-Entry-Box).
59. Transcript, TGA Meeting in Grand Junction, Colorado, September 17, 1934, at 2 (FRC-DPL).
60. Department of the Interior Press Release, January 13, 1936 (NA-DNV, 49-39-140).
61. Transcript, TGA Hearing in Cody, Wyoming, August 2, 1934, at 17 (NA-DNV, 49-45-171).
62. Transcript, TGA Meeting in Grand Junction, Colorado, September 17, 1934, at 11 (FRC-DPL).
63. For examples regarding the McCarran amendment, see his remarks during TGA setup meetings in early 1935. Transcript, Statewide Meeting for Nevada, January 24, 1935, at 111–16 (FRC-DPL); Transcript, Statewide Meeting for Arizona, January 28, 1935, at 82 (FRC-DPL). The grazing privileges directive is located in a letter from FRC to Archie Ryan, September 16, 1935 (NA-DNV, 49-39-137).
64. Letter, Carpenter to Solicitor Nathan Margold, September 12, 1934 (FRC-DPL). Carpenter went so far as to seek informal confirmation of his view from Assistant Attorney General Harold M. Stephens, who agreed with his assessment completely. Letter, Harold Stephens to FRC, July 6, 1935 (FRC-DPL).
65. Letter, Carpenter to Solicitor Nathan Margold, September 12, 1934 (FRC-DPL).
66. Letter, Margold to Carpenter, September 21, 1934 (FRC-DPL). The solicitor's solution was that licenses could be issued under the broad discretion given to the secretary in Section 2 of the law, which stated, "The Secretary of the Interior shall ... make such rules and regulations ... and do any and all things necessary to accomplish the purposes of this Act ..." 48 Stat. 1269, Sec. 2. For approval of the rules, see Memo, Division of Grazing Acting Director John F. Deeds to Secretary Ickes, approved May 29, 1935 (NA-

MD, 48-749A-788), and Thomas E. Buckman, Setting Up Taylor Grazing Districts in Nevada, Bulletin 77, Agricultural Extension Service, University of Nevada, June 30, 1937, at 19 (AHC, 675-7).

67. Letter, Carpenter to Ryan, March 22, 1935 (NA-DNV, 49-39-137).

68. Letter, F. E. Mollin to American National Live Stock Association Public Lands Committee, May 13, 1935 (Frank Delaney Collection, Box 194, Folder 3,University of Colorado at Boulder, Norlin Library Archives; hereafter cited as FD, Box-Folder).

69. An early copy of the speech by Interior official Theodore Walters at the January 1936 district advisors' conference includes handwritten corrections of the terms *permit license,* and *lease,* leading Walters to make the correct statement at the actual conference: "A permit is an asset, different in character from a license and would justify a larger fee." Department of Interior Memorandum for the Press regarding Walters's speech, January 13, 1936, at 8 (NA-DNV, 49-39-140); and Transcript, District Advisors' Convention, Salt Lake City, January 13–14, 1936, at A7 (FRC-DPL).

70. Reported in the *National Wool Grower,* May 1935, at 7. In a May 1935 meeting with stockmen, Carpenter mentioned "90 days" to "next year" as being a possible time period for the change. By August of the same year, he was not mentioning any specific timelines. See Thomas E. Buckman, Setting Up Taylor Grazing Districts in Nevada, Bulletin 77, Agricultural Extension Service, University of Nevada, June 30, 1937, at 19 (AHC, 675-7), and Transcript, Wyoming Wool Growers 1935 Convention, Laramie, Wyoming, July 30–August 1, 1935, at 170 (AHC, 1350-5).

71. Letter, FRC to all Regional Graziers, March 1, 1937 (NA-DNV, 49-45-145).

72. Transcript, TGA Hearing in Cody, Wyoming, August 2, 1934, at 70 (NA-DNV, 49-45-171); Resolutions, New Mexico Cattle Growers Association, December 28, 1934 (NA-DC, 49-3-13); Resolutions, Colorado Stock Growers and Feeders Association, February 11, 1935 (NA-DC, 49-3-13).

73. TGA Hearing in Cody, Wyoming, August 2, 1934, at 70. See also Transcript, Conference on TGA at Boise, Idaho, July 26–27, 1934, at 10b (NA-DNV, 49-45-171).

74. Letter, Archie Ryan to Lona Thomason, August 14, 1935 (NA-DNV 49-45-163).

75. Transcript, District Advisors' Convention, Salt Lake City, January 13–14, 1936, at A37 (FRC-DPL).

76. Testimony of Interior representative Rufus Poole, Hearings before House Public Lands Committee on H.R. 3019, March 11, 1935, at 117. Carpenter tried to discourage interest in leasing public land under Section 15 of the act by stressing the relative security of fenced allotments within grazing districts under Section 3. Transcript, Statewide Meeting for Oregon, December 15, 1934, at 52 (FRC-DPL).

77. Emphasis added. Transcript, Statewide Meeting for Arizona, January 28, 1935, at 67 (FRC-DPL).

78. Carpenter proclaimed that they will "practically perform a marriage ceremony when [they] give the term permit" between the commensurate property and the public range. See Transcript, Second Annual District Advisrs' Conference, Salt Lake City, December 9–11, 1936, at 28 (NA-DNV, 49-44-144).

79. For example, see reply by Oscar Chapman to E. D. Blodgett, United States Bank of Grand Junction, July 26, 1935 (NA-MD, 48-749A-791); or Letter, Conrad Ball, Federal Intermediate Credit Bank of Wichita, to Division of Grazing, December 31, 1935 (NA-DNV, 49-45-163).

80. For example, see comments of Rufus Poole, Conference on TGA at Boise, Idaho, July 26–27, 1934, at 9 (NA-DNV, 49-45-171); comments of Senator Joseph O'Mahoney of Wyoming, TGA Hearing in Cody, Wyoming, August 2, 1934, at 1 (NA-DNV, 49-45-171).

81. Speech by William Grounds, Prescott, Arizona, September 19, 1934 (AHC, 3024-1).
82. Transcript, District Advisors' Convention, Salt Lake City, January 13–14, 1936, at A29 (FRC-DPL).
83. Statement of Oliver Lee, Transcript, District Advisors' Convention, Salt Lake City, January 13–14, 1936, at B9 (FRC-DPL).
84. For the priority argument, see letter, Jess Hawley to FRC, January 21, 1935 (NA-DNV, 49-45-145). For an example of the argument against new beginners, see testimony, TGA Hearing in Cody, Wyoming, August 2, 1934, at 34–35 (NA-DNV, 49-45-171).
85. Transcript, District Advisors' Convention, Salt Lake City, January 13–14, 1936, at A33 (FRC-DPL).
86. Transcript, Statewide Meeting for Nevada, January 24, 1935, at 78 (FRC-DPL).
87. Division of Grazing Circular No. 2, issued May 31, 1935 (NA-DNV, 49-39-139).
88. See Memo, Acting Solicitor Charles Fahy to Secretary Ickes, May 2, 1935 (NA-MD, 48-749A-792).
89. Buckman (1937), at 24–25.
90. Letter, Archie Ryan to Jack Deeds, August 14, 1935 (NA-DNV, 49-39-137).
91. For example, see Letter, Archie Ryan, Division of Grazing, to A. M. Myrup, August 31, 1935 (NA-DNV, 49-45-164).
92. See Letter, G. M. Kerr, Division of Grazing, to John H. Johnson, October 3, 1935 (NA-DNV, 49-45-165); or Letter, J. Q. Peterson, Division of Grazing, to W. G. King, October 5, 1935 (NA-DNV, 49-45-164).
93. Carpenter made the point in his typically witty manner: "If we try to get these districts under regulations as hastily as many people think we should we will hamper and hamstring the livestock and hammer the heads of the operators unmercifully. If we go at it slow we will continue to hammer the public domain. Well, as the public domain range is less articulate than the stockmen, we have chosen to hammer the public domain." Transcript, Statewide Meeting for Oregon, December 15, 1934, at 31 (FRC-DPL).
94. "Landowners," declared the director, "heretofore not using public domain will not be excluded permanently." Telegram, FRC to Sid Smith, January 11, 1935 (NA-DNV, 49-45-167).
95. Transcript, Statewide Meeting for Arizona, January 28, 1935, at 79 (FRC-DPL). In a similar vein, see also Transcript, Statewide Meeting for Idaho, December 17, 1934, at 70 (FRC-DPL); Transcript, Statewide Meeting for Nevada, January 24, 1935, at 93 (FRC-DPL).
96. Transcript, Statewide Meeting for Nevada, January 24, 1935, at 81 (FRC-DPL).
97. For example, see Letter, Archie Ryan to FRC, May 27, 1935 (NA-DNV, 49-39-137); Carpenter's presentation to Idaho stockmen, Transcript, Statewide Meeting for Idaho, December 17, 1934, at 43 (FRC-DPL); or Thomas Havell's statement, transcript, TGA Hearing in Cody, Wyoming, August 2, 1934, at 71 (NA-DNV, 49-45-171).
98. For example, see Plans for Government Control of Grazing on the Public Domain, *National Wool Grower*, August 1934, at 14; or Transcript, Conference on TGA at Boise, Idaho, July 26–27, 1934, at 11b, 15b (NA-DNV, 49-45-171). Interior staff and even large operators seemed at first to support, or at least expect, these distributions for new entrants. See Letter, Frank Delaney to F. E. Mollin, Secretary, ANLA, August 14, 1934 (FD, 194-2); Carpenter's remarks on the subject in Transcript, Statewide Meeting for Oregon, December 15, 1934, at 70 (FRC-DPL); or testimony of Assistant. Solicitor Rufus Poole, House Hearings on H.R. 3019, March 1, 1935, at 22. A draft version of the first set of range rules in 1935 included a minimum (or protective) limit of 150 cattle or 750 sheep

below which no graziers would be reduced until all larger operators had reached the same level. See Draft Circulars for 1935, Division of Grazing (NA-DNV, 49-40-141).

99. For example, see Letter, F.E. Mollin, ANLA, to Public Lands Committee, September 8, 1934 (FD, 194-2).

100. Buckman (1937), at 22.

101. Transcript, Wyoming Wool Growers Convention, Laramie, Wyoming, July 30–August 1, 1935, at 128 (AHC, 1350-5).

102. Transcript, Statewide Meeting for Arizona, January 28, 1935, at 28-30 (FRC-DPL). The point is echoed by Colorado attorney and rancher Frank Delaney, who frequently provided the director with advice, in a letter to Carpenter dated September 11, 1935: "the one who goes to this additional expense and trouble [owning commensurate property] should be rewarded for his prudence, his toil, and the money he invests in the community" (FD, 194-3).

103. For specific "common good" rhetoric in this vein, see Press Release, text of speech by Secretary Ickes at Denver TGA Conference, February 12, 1935, at 18 (NA-DNV, 49-39-139).

104. Transcript, District Advisors' Convention, Salt Lake City, January 13–14, 1936, at A32 (FRC-DPL).

105. Transcript, Statewide Meeting for Idaho, December 17, 1934, at 33 (FRC-DPL).

106. Press Release, text of speech by Secretary Ickes at Denver TGA Conference, February 12, 1935, at 18, 22 (NA-DNV, 49-39-139); see also Transcript, Conference on TGA at Boise, Idaho, July 26–27, 1934, at 2a (NA-DNV, 49-45-171).

107. Carpenter recalled puzzling over the idea so much even at home that one day his son greeted him with the question, "Good morning Dad, how's your commensurability?" Transcript, Statewide Meeting for Arizona, January 28, 1935, at 80 (FRC-DPL).

108. Division of Grazing Circular No. 2, issued May 31, 1935 (NA-DNV, 49-39-139).

109. Carpenter, Establishing Management under the Taylor Grazing Act, speech delivered at Montana State College, January 8, 1962, at 10 (FRC-DPL).

110. A Division of Grazing staff conference in September 1935, for example, was devoted entirely to defining commensurate property. See Note, FRC to Jack Deeds, September 2, 1935 (NA-DNV, 49-39-137).

111. Initially uncertain, the department declared in 1935 that leasing private land would meet the commensurate property requirement. Although the 1935 rules accepted leasing, Carpenter continued to discuss equitable arguments on both sides of the issue with his staff. On the one hand, the division wanted to reward the settled user of the public domain who operated a geographically stable year-round livestock operation. From this perspective, whether he leased or owned his land was immaterial. On the other hand, the division worried that transient operators would now lease private lands with attached public domain rights, thereby continuing to secure public range forage without becoming settled members of a local community. The danger that leasing would serve as a means of evading the division's goal of encouraging property-owning stock operations made it remain controversial into future years. See Transcript, TGA Hearing in Cody, Wyoming, August 2, 1934, at 14, 37 (NA-DNV, 49-45-171); Transcript, Statewide Meeting for Arizona, January 28, 1935, at 35 (FRC-DPL); Note, FRC to Jack Deeds, September 2, 1935 (NA-DNV, 49-39-137).

112. Because buying hay was easy for an itinerant herder and had nothing itself to do with stable patterns of land tenure, the division initially resisted recognizing it as a valid form of commensurate property. As Carpenter noted at a meeting in 1934, "There is a certain line

of demarcation between a contract [to buy hay] and a lease. One runs into a tenure of land and the other does not." Transcript, Statewide Meeting for Oregon, December 15, 1934, at 62 (FRC-DPL). Yet the desire to exclude purchased hay as commensurate property created problems for some long-time prior users of the range. In Idaho and Nevada, for example, many range users with small private setups typically bought winter feed from farmers who themselves had no previous use of the public grazing lands. Strict adherence to the commensurate property rule in these cases would have given the range rights to the farmers, not the ranchers. In these cases, Carpenter realized that purchased hay might have to serve as commensurate property if prior users were to be given any access to the public domain. Transcript, Statewide Meeting for Nevada, January 24, 1935, at 49–50 (FRC-DPL). The possibility of recognizing purchased hay as a discounted form of commensurate property was even considered as one solution. Transcript, Statewide Meeting for Nevada, January 24, 1935, at 119 (FRC-DPL). The division's 1935 rules permitted the use of purchased feed as commensurate property; nevertheless, treatment of hay as a form of commensurate property varied by district and remained a controversial topic into 1936 and beyond. Division of Grazing Circular No. 2, issued May 31, 1935 (NA-DNV, 49-39-139).

113. Letter, Acting Director N.F. Waddell to Assistant Secretary Oscar Chapman, September 11, 1934 (NA-MD, 48-768-10). At an early hearing in Colorado, Carpenter said that one lawyer had given him a brief with 57 definitions of the term *near*. Carpenter initially wanted to give a definitive, "John Marshall" decision on what exactly the term meant under the rules, but he soon took the advice of Forest Service employees who told him, "when you do not have to decide anything right away, just put it off." Transcript, Statewide Meeting for Oregon, December 15, 1934, at 67 (FRC-DPL).

114. For example, see Transcript, Statewide Meeting for Oregon, December 15, 1934, at 68 (FRC-DPL); see also Transcript, Conference on TGA at Boise, Idaho, July 26–27, 1934, at 4b (NA-DNV, 49-45-171).

115. The 1935 rules went on to favor properties closer to the public domain when priority rules failed to narrow the pool of applicants adequately. Division of Grazing Circular No. 2, issued May 31, 1935 (NA-DNV, 49-39-139). Given the strict priority rules used in many districts, however, it appears that location rarely made much difference in actual allocations at this time.

116. Letter, FRC to Archie Ryan, September 24, 1935 (NA-DNV, 49-39-137).

117. Transcript, District Advisors' Convention, Salt Lake City, January 13–14, 1936, at B35 (FRC-DPL).

118. The uncommon emphasis on local input in the rulemaking process was noted by the press. For example, see Stockmen to Discuss Taylor Grazing Act at Salt Lake, *Denver Post*, December 23, 1935.

119. Transcript, District Advisors' Convention, Salt Lake City, January 13–14, 1936, at A26 (FRC-DPL).

120. Transcript, District Advisors' Convention, Salt Lake City, January 13–14, 1936, at A35 (FRC-DPL).

121. Transcript, District Advisors' Convention, Salt Lake City, January 13–14, 1936, at A27 (FRC-DPL).

122. Circular Letter No. 11, Nevada Extension Service, January 17, 1936 (NA-DNV, 49-45-163).

123. Transcript, District Advisors' Convention, Salt Lake City, January 13–14, 1936, at B15 (FRC-DPL).

124. Rules for Administration of Grazing Districts, approved March 2, 1936, at 1 (NA-DNV, 49-33-129).

125. *The Grazing Bulletin* I(2), June 1936, at 3 (NA-DNV, 49-33-129). Letter, Secretary Ickes to Senator McCarran, March 17, 1936 (NA-MD, 48-749A-788).

126. Annual Report of the Secretary of the Interior, 1938, at 107.

127. Letter, Regional Grazier J. E. Stablein to FRC, March 31, 1937 (NA-DC, 49-3-13); Letter, Jerrie Lee, Arizona Wool Growers Association to Secretary Ickes, August 2, 1938 (NA-DC, 49-3-12).

128. Legal Problems in Grazing Regulations, Office of the Secretary of the Interior, 1936 (NA-DC, 49-3-18); for similar sentiments, see speech by FRC in Transcript, Wyoming Wool Growers Conference, July 29, 1936, at 107 (AHC, 1350-5), or The Government and Our Grazing Lands, *National Wool Grower*, September 1936 (AHC, 7889-191).

129. *The Grazing Bulletin* I(5), October 1937, at 13 (NA-DNV, 49-33-129).

130. Proposed Rules for Administration of Grazing Districts, November 4, 1937, at 5–6 (NA-DNV, 49-45-147).

131. For an example of a favorable review, see Letter, Regional Grazier Marvin Klemme to FRC, September 23, 1937 (NA-DNV, 49-45-147); for opposed, see Letter, Regional Grazier Lester Brooks to FRC, September 27, 1937 (NA-DNV, 49-45-147); for uncertain, see Letter, Delaney to Regional Grazier C. F. Moore, September 22, 1937 (FD, 194-5). The final version of the rule is in Federal Range Code, Sec. 7(a), 7(b) (NA-DC, 49-3-9).

132. See *Cow Country* 64(1), June 22, 1936 (AHC, 14-284); for the practical side, see Memo, A. D. Ryan to Director, Division of Grazing, September 19, 1939, asserting that New Mexico licensees considered their allotments as vested rights (NA-DC, 49-4-5).

133. "You understand that once a permit issues under the McCarran amendment it is there for all time." Letter, FRC to Delaney, September 28, 1937 (FD, 194-5).

134. Memo, Assistant Solicitor Neil Stull to Solicitor Margold, October 1, 1938, at 12 (NA-MD, 48-835-7).

135. For example, see speech by FRC, Wyoming Wool Growers Convention, July 29, 1936, at 101 (AHC, 1350-5).

136. Small users in the Southwest, especially New Mexico, were vocal about more recognition of prior use. See petition to Secretary Ickes from stockmen in New Mexico Grazing District 5, March 1936 (NA-DC, 49-4-5). Landowners and others in Oregon took the opposite view—that priority was being used far too much as an allocation principle under the law. See open letter to FRC from Western Livestock Producers Association, October 4, 1936 (NA-DNV, 49-45-162).

137. For example, Letter, G. M. Kerr to FRC, January 28, 1937 (NA-DC, 49-3-9); Resolution 34, National Wool Growers Convention, January 26–28, 1938 (NA-DC, 49-3-14).

138. Field Bulletin, FRC to Division of Grazing Staff, August 18, 1936 (NA-DNV, 49-45-145).

139. Letter, FRC to Robert Stanfield, Western Livestock Producers Association, October 16, 1936, at 3 (NA-DNV, 49-45-162). See also Carpenter's speech, Wyoming Wool Growers Convention, July 29, 1936, at 101 (AHC, 1350-5). Carpenter still expressed hope as late as November 1936 that the importance of the priority rules would "fade out considerably" by the time permits were issued on the basis of commensurate property land use. See Letter, FRC to C. F. Moore, November 13, 1936 (NA-DNV, 49-45-156).

140. For example, see Livingston's testimony in partial transcript of hearing before District 6 advisory board, May 4, 1936, at 3 (NA-DNV, 49-270-2).

141. Letter, FRC to Terrett, August 17, 1936 (NA-DNV, 49-39-137).

142. For example, see the remarkable exchange of letters between Assistant Director Julian Terrett, Frank Delaney, and FRC, July 29, August 7, and August 17, 1936 (NA-DC, 49-6-6 and NA-DNV, 49-39-137).

143. According to Carpenter, the active interest of the Solicitor's Office in priority late in 1936 stemmed directly from its consideration of the Livingston case. Note, FRC to Delaney, 1937 (FD, 194-5).

144. Assistant Secretary Walters, for example, found the language on priority "indefinite and unsatisfactory." Memo to Secretary Ickes, February 29, 1936 (NA-MD, 48-777-9). Carpenter, on the other hand, had found the new definition of priority in the rules "the most advanced step in the entire regulations." Letter, FRC to Archie Ryan, March 4, 1936 (NA-DNV, 49-39-137). See also Memo, Archie Ryan to Field Employees, March 12, 1936 (NA-DNV, 49-45-163).

145. Memos, Solicitor's Office, October 23, 1936 and December 1936 (NA-MD, 48-835-8); Note, FRC to Julian Terrett, November 27, 1936 (NA-DNV, 49-41-142).

146. Transcript, Second Annual District Advisors' Conference, Salt Lake City, December 9–11, 1936, at 19 (NA-DNV, 49-44-144).

147. Transcript, Second Annual District Advisors' Conference, Salt Lake City, December 9–11, 1936, at 25 (NA-DNV, 49-44-144).

148. That is, the conference passed a resolution stating that "determination of the priority rule be left to the Advisory Boards of each district of the several states." Transcript, Second Annual District Advisors' Conference, Salt Lake City, December 9–11, 1936, at 56 (NA-DNV, 49-44-144).

149. Program, Second Annual Conference of District Advisors, Salt Lake City, December 9, 1936 (NA-DNV, 49-45-145).

150. Ideas ranged from omitting priority altogether, to qualifying anyone with even a day's use before the passage of the TGA, to requiring a full year of prior use. Assistant Solicitor Rufus Poole argued that the one-day rule was "manifestly more fair and more reasonable" than the alternatives, while Assistant Solicitor Leland Graham, who remained extensively involved in grazing issues during 1937, favored the one-year rule. See Memo, Poole to Margold and Walters, January 27, 1937 (NA-MD, 48-749B-3250), and Memo, Leland Graham, January 18, 1937 (NA-MD, 48-834-1).

151. Memo, Margold to Ickes, January 21, 1937, at 5 (NA-DNV, 49-45-147).

152. Division of Grazing staff member A. D. Molohon sarcastically told Carpenter that the new one-year-in-five rule made "quite a hit with the boys at the [stockmen's] convention he was attending." Letter, Molohon to FRC, January 28, 1937 (NA-DC, 49-3-9).

153. Note, FRC to Delaney, March 12, 1937 (FD, 194-5).

154. Appeal from the Division of Grazing, Case of Joseph F. Livingston, March 29, 1937 (NA-DNV, 49-45-156). Ironically, the secretary opted to dismiss Livingston's 1935 application as "moot" because the season was now long since finished, even though the endless delays hearing the sheepman's appeals were entirely the fault of the Interior Department staff.

155. A few even supported it enthusiastically. For example, see letter, Regional Grazier Marvin Klemme to FRC, February 24, 1937 (NA-DNV, 49-45-162).

156. Note, FRC to Delaney, 1937 (FD, 194-5).

157. Letter, FRC to Delaney, April 8, 1937 (DPL-FRC); Note, FRC to Delaney, 1937 (FD, 194-5).

158. Memo, FRC to Ickes, March 13, 1937; Note, FRC to Delaney, April 1, 1937 (FD, 194-5).

159. "We are dealing with a paranoiac and a lot of kids and maybe the sooner the West finds it out the better for everyone. I have not given up hope ... but I'm damn near to the end of my rope." Note, FRC to Delaney, March 12, 1937 (FD, 194-5).

160. Memo to the Grazing Policy Committee, April 2, 1937 (NA-DC, 49-3-2).

161. Minutes of the Grazing Policy Committee Meeting, May 25, 1937 (NA-DC, 49-3-2).

162. For example, Transcript, Wyoming Wool Growers Meeting, Kemmerer, Wyoming, August 4–6, 1937, at 94-5 (AHC, 1350-5).

163. Initially specified in the September 27, 1937, version of the Proposed Rules for Administration of Grazing Districts (NA-DNV, 49-45-147).

164. *The Grazing Bulletin* 2(1), March 1938, at 4 (NA-DNV, 49-33-129); Range Users Okeh [*sic*] New Grazing Rules, *Tribune*, December 1, 1937 (AHC, Taylor Grazing Act Subject File #T212ga).

165. Statement of FRC, Memorandum for the Press, March 16, 1938, at 5 (NA-MD, 48-835-5).

166. Class 1 applicants remained those who owned property "dependent by location and use," meaning property that was near the public range and had prior use associated with it. Federal Range Code, March 16, 1938, at 4 (NA-DC, 49-3-9).

167. Federal Range Code, March 16, 1938, at 2 (NA-DC, 49-3-9).

168. Eventually, the sheepherder gave up the fight and bought new property in Colorado with an extensive history of prior use. Ironically, he purchased the property from his long-time legal nemesis, cattleman Frank Delaney.

169. The issue was raised for the solicitor by Assistant Interior Secretary Oscar Chapman, Memo to the Solicitor, June 16, 1938 (NA-MD, 48-835-6).

170. Amendments to the Federal Range Code as submitted to the secretary, August 15, 1938, at 1 (NA-MD, 48-835-6).

171. For example, see Transcript, Second Annual District Advisors' Conference, Salt Lake City, December 9–11, 1936, at 15 (NA-DNV, 49-44-144); or Letter, V. F. Christensen to FRC, July 1, 1938 (NA-DNV, 49-44-144).

172. Letter, FRC to J. S. Westwood & Sons, February 23, 1937 (NA-DNV, 49-45-159)

173. Report to Bradley B. Smith, Division of Investigations, July 20, 1938, at 5–6 (NA-DC, 49-3-2). Carpenter later maintained that operators on the public domain in general did not want any sort of maximum limit, preferring instead the chance to become a big operator someday. Interview with FRC, May 21, 1955, at 3 (Colorado Historical Society, MSS#106, Box 1).

174. Letter, FRC to C. F. Moore, May 6, 1938, as reprinted in Letter, FRC to Secretary of the Interior, August 18, 1938 (NA-DC, 49-3-2).

175. See, for example, Letter Acting Secretary W.C. Mendenhall to Senator Charles McNary, October 30, 1936 (NA-MD, 48-749A-788); Transcript, Second Annual District Advisors' Conference, Salt Lake City, December 9–11, 1936, at 13 (NA-DNV, 49-44-144); or summary, Meeting of the Colorado District Boards, October 28, 1937, at 6 (NA-DC, 49-3-10).

176. Emphasis added. Report, Summary of Accomplishments under the Act of June 26, 1934, submitted to the secretary August 18, 1937, at 5 (NA-DNV, 49-42-143).

177. Letter, Marvin Klemme to FRC, October 12, 1936 (NA-DNV, 49-39-137).

178. Letter, Charles Redd to FRC, October 1, 1937 (NA-DNV, 49-45-147).

179. See series of articles, for example, from the March 1938 issues of the *Idaho Statesman* condemning the potential allocation of rights to commensurate property owners without prior use (NA-MD, 48-749B-3238).

180. Letter, FRC to Carl Leonard, January 29, 1938 (NA-DC, 49-3-13). The division also cited keeping new operators off the public domain as an achievement both to existing stockmen and to the Office of the Secretary. See statements of FRC, Transcript, Second Annual

District Advisors' Conference, Salt Lake City, December 9–11, 1936, at 10 (NA-DNV, 49-44-144); and Memorandum for the First Assistant Secretary, November 30, 1936, at 14 (NA-DC, 49-3-18).

181. Letter, FRC to G. M. Kerr, March 23, 1937 (NA-DNV, 49-45-148).

182. Letter, Regional Grazier J. E. Stablein to FRC, September 9, 1936 (NA-DNV, 49-45-160); Letter, Stablein to A. D. Brownfield, January 22, 1937 (NA-DNV, 49-45-161). In some cases, the arguments of the small stockmen were grounded in priority terms as well.

183. The biggest case of this type was the appeal of homesteader Arthur Pue, which stretched from 1936 into 1938. After a fascinating series of legal twists and turns, Pue eventually won a grazing permit, despite a lack of priority on the land in question. See Decision of the Director of Grazing, July 19, 1938 (NA-DC, 49-6-17).

184. The bill was S. 3934, 75th Congress, 3rd Session. See Letter, E. K. Burlew to Senator Alva Adams, June 11, 1938 (NA-DNV, 49-40-141).

185. Carpenter wrote to Delaney on more than one occasion expressing support for a tough rule against leasing. See Letters, FRC to Delaney, March 1 or September 28, 1937; in the latter, Carpenter contended that "the change now being considered is to bar all leased lands from ever getting a permit. ... " (FD, 104-5). The distinction between leases on private and public land remained critical to this debate. See Letter, Regional Grazier Harold Burback to FRC, September 13, 1937 (NA-DC, 49-3-13); New Mexico Advisory Board Resolution II, August 17, 1936 (NA-DNV, 49-45-160). In a more extreme example, cattleman Delaney recommended not recognizing any leased lands as being equally qualified to those owned in fee. Letter, Delaney to F. E. Mollin, February 20, 1937; Letter, Delaney to Secretary Ickes, February 23, 1937 (FD, 194-5).

186. Federal Range Code, March 16, 1938, at 2, 5 (NA-DC, 49-3-9) and Memo to the Grazing Policy Committee, December 9, 1937, at 2 (NA-DC, 49-3-2). Although Forest Service permits were not a form of commensurate property, via a rule known as the Oregon compromise" they could be used as a supplement of sorts for operators owning some, but not enough, private commensurate property for their herds.

187. Although purchased hay was recognized in the 1935 rules, by late 1936 it counted as commensurate property only with the explicit waiver of range privileges by the landowners who grew the feed. Memorandum from FRC re: Staff Conference, Salt Lake City, October 1–3, 1936, at 4 (NA-DNV, 49-45-147). In the 1938 range code, purchased hay had the same status as Forest Service permits.

188. Federal Range Code, March 16, 1938, at 3 (NA-DC, 49-3-9).

189. Letter, L. R. Brooks to FRC, September 25, 1937 (NA-DNV, 49-45-147).

190. Note, FRC to G. M. Kerr, March 29, 1937 (NA-DNV, 49-45-147).

191. The code was not fully revised again until 1942, and those changes remained largely faithful to the 1938 act.

Chapter 5: Allocating Greenhouse Gas Emissions, 1992–1997

1. At least for those who prefer a definition of the commons that distinguishes it from an open-access resource subject to unlimited use, as was discussed in Chapter 1 (Bromley 1991). In practice, common property entails significant restrictions on resource use, whereas resources in the so-called global commons remains substantially open to all comers. Authors who use the global commons language recognize the open-access qual-

ity of these resources, of course, but employ the phrase anyway. In this respect they risk perpetuating the overextension of Garrett Hardin's (1968) Tragedy of the Commons model by not properly distinguishing a commons from an open-access resource. A better, or at least more semantically consistent, approach might be to refer to such resources as extranational, to emphasize their location outside any state's sovereignty.

2. For a brief discussion of the literature concerning the important role of Hegelian ideas in American politics and policy, see Raymond and Fairfax (1999), at 666–69.
3. See http://unfccc.int/resource/conv/ratlist.pdf (accessed April 3, 2003).
4. Report of the Ad Hoc Group on the Berlin Mandate (AGBM) on the Work of Its First Session, FCCC/AGBM/1995/2, September 28, 1995, at 5.
5. Report of the Ad Hoc Group on the Berlin Mandate (AGBM) on the Work of Its First Session, FCCC/AGBM/1995/2, September 28, 1995, at 3.
6. Report of the AGBM on the Work of Its Eighth Session, FCCC/AGBM/1997/8, November 19, 1997, at 3. For a sampling of other general appeals for an "equitable" distribution of emissions, see Comments from Parties, Note by the Secretariat, FCCC/AGBM/1996/Misc.1/Add.3, June 27, 1996, at 4 (Japan seeking an "equitable and efficient distribution of future effort [regarding QELROs] among the Annex I Parties"); or comments of various nations reported in *Earth Negotiations Bulletin* 12:27, March 11, 1996, at 6–8.
7. See, for instance, Review of Possible Indicators to Define Criteria for Differentiation among Annex I Parties: Note by the Secretariat, FCCC/AGBM/1996/7, June 21, 1996, at 3, and Comments of Norway, at 2; *Earth Negotiations Bulletin* 12:45, March 10, 1997, at 4; Report of the AGBM on the Work of Its Second Session, FCCC/AGBM/1995/7, August 29, 1997, at 9.
8. For example, see Comments of Australia in Proposals from Parties, Note by the Secretariat, FCCC/AGBM/1996/Misc.2/Add.2, November 15, 1996, at 6 and 14.
9. *Earth Negotiations Bulletin*, 12:38, July 22, 1996, at 12.
10. *Earth Negotiations Bulletin*, 12:45, March 10, 1997, at 1.
11. The United States specifically references the acid rain program in a "non-paper" submitted to the AGBM. See Proposals by Parties, Note from the Secretariat, FCCC/AGBM/1996/Misc.2/Add.4, December 10, 1996, at 31.
12. *Earth Negotiations Bulletin*, 12:27, March 11, 1996, at 6–7.
13. As argued by New Zealand, Proposals by Parties, Note by the Secretariat, FCCC/AGBM/1996/Misc.2/Add.4, December 10, 1996, at 15.
14. Germany's submission on this point expresses the view well: "Emission reduction should basically be carried out within each Party's own territory." Framework Compilation of Proposals from Parties for the Elements of a Protocol or Another Legal Instrument, Note by the Chairman, FCCC/AGBM/1997/2, January 31, 1997, at 45.
15. *Earth Negotiations Bulletin* 12:76, December 13, 1997, at 11. For examples of requests to delete emissions trading from the AGBM process, see also *Earth Negotiations Bulletin*, 12:38, July 22, 1996, at 8, and *Earth Negotiations Bulletin* 12:45, March 10, 1997, at 4. See also Proposals from Parties, Note by the Secretariat, FCCC/AGBM/1997/Misc.1/Add.8, October 30, 1997, at 6.
16. See Report of the AGBM on the Work of Its Fourth Session, FCCC/AGBM/1996/8, October 7, 1996, at 7; or discussion of legally binding commitments in Possible Features of a Protocol or Another Legal Instrument, FCCC/AGBM/1996/6, May 21, 1996, at 6–8.
17. Report of the AGBM on the Work of Its Fourth Session, FCCC/AGBM/1996/8, October 7, 1996, at 15.

18. Comments from Parties, Note from the Secretariat, FCCC/AGBM/1996/Misc.2, May 17, 1996, at 37.
19. Synthesis of Proposal by Parties, Note by the Chairman, FCCC/AGBM/1996/10, November 19, 1996, at 7.
20. See Proposals from Parties, Note by the Secretariat, FCCC/AGBM/1996/Misc.2/Add.2, November 15, 1996, at 50; Proposals from Parties, Note by the Secretariat, FCCC/AGBM/1996/Misc.2/Add.4, December 10, 1996, at 27. The United States was joined by selected other nations in this argument, including Canada. See Additional Proposals from Parties, Note by the Secretariat, FCCC/AGBM/1997/Misc.1/Add.2, May 5, 1997, at 15–16.
21. Framework Compilation of Proposals from Parties for the Elements of a Protocol or Another Legal Instrument, FCCC/AGBM/1997/2/Add.1, February 26, 1997, at 5; Proposals for a Protocol or Another Legal Instrument, Negotiating Text by the Chairman, FCCC/AGBM/1997/3/Add.1, April 22, 1997, at 6.
22. The closest thing to a prior use argument that emerges from the record is a sentence offered by the United Kingdom in defense of flat rate reductions from the status quo. In addition to being practical and simple, such an approach "is also equitable in the sense that each Party's efforts are measured against its historic levels of emissions." This argument could be interpreted as a quasi-Lockean appeal to entitlements based on prior use, but it is hardly a ringing endorsement of that idea, as was commonplace in the domestic cases. Nor is it much echoed by other statements in the record for this case. See Proposals from Parties, Note by the Secretariat, FCCC/AGBM/1996/Misc.2, May 17, 1996, at 37.
23. For one example, see Proposals from Parties, Note by the Secretariat, FCCC/AGBM/1996/Misc.2/Add.2, November 15, 1996, at 50.
24. Proposals from Parties, Note by the Secretariat, FCCC/AGBM/1996/Misc.2/Add.4, December 10, 1996, at 29. See also Proposals from Parties, Note by the Secretariat, FCCC/AGBM/1996/Misc.2/Add.2, November 15, 1996, at 51.
25. As reiterated in Proposals from Parties, Note by the Secretariat, FCCC/AGBM/1996/Misc.2, May 17, 1996, at 5.
26. Additional Proposals from Parties, Note by the Secretariat, FCCC/AGBM/1997/Misc.1/Add.2, May 5, 1997, at 60.
27. *Earth Negotiations Bulletin*, 12:45, March 10, 1997, at 1.
28. Proposals from Parties, Note by the Secretariat, FCCC/AGBM/1996/Misc.2, May 17, 1996, at 32.
29. Proposals from Parties, Note by the Secretariat, FCCC/AGBM/1996/Misc.2, May 17, 1996, at 32.
30. As noted by the United Kingdom, in Proposals from Parties, Note by the Secretariat, FCCC/AGBM/1996/Misc.2, May 17, 1996, at 37.
31. In Submission of Norway, Proposals from Parties, Note by the Secretariat, FCCC/AGBM/1996/Misc.2/Add.2, November 15, 1996, at 27.
32. For instance, see comments reported in *Earth Negotiations Bulletin* 12:27, March 11, 1996, at 5–7.
33. For example, see Proposals from Parties, Note by the Secretariat, FCCC/AGBM/1996/Misc.2/Add.2, November 15, 1996, at 20 and 30.
34. Proposals from Parties, Note by the Secretariat, FCCC/AGBM/1996/Misc.2/Add.4, December 10, 1996, at 17. The 1994 Sulphur Protocol to the Convention on Long-Range Transboundary Air Pollution is also mentioned in the AGBM process as a prior example

of using the initial allocation to help equalize the marginal costs of reducing emissions. See Review of Relevant Conventions and Other Legal Instruments, Note by the Secretariat, FCCC/AGBM/1996/6, May 21, 1996, at 13.

35. Proposals from Parties, Note by the Secretariat, FCCC/AGBM/1996/Misc.2/Add.2, November 15, 1996, at 30–31.

36. Review of Relevant Conventions and Other Legal Instruments, Note by the Secretariat, FCCC/AGBM/1996/6, May 21, 1996, at 11.

37. Review of Possible Indicators to Define Criteria for Differentiation among Annex I Parties, Note by the Secretariat, FCCC/AGBM/1996/7, June 21, 1996, at 10.

38. For example, see Consolidated Negotiating Text by the Chairman, FCCC/AGBM/1997/7, October 13, 1997, at 25.

39. For one such reference to per capita emissions in developing nations as being "relatively low," see Proposals from Parties, Note by the Secretariat, FCCC/AGBM/1996/Misc.2, May 17, 1996, at 3.

40. Proposals from Parties, Note by the Secretariat, FCCC/AGBM/1996/Misc.2/Add.4, December 10, 1996, at 12.

41. Comments from Parties, Note by the Secretariat, FCCC/AGBM/1995/Misc.1/Add.2, October 25, 1995, at 13, and Framework Compilation of Proposals from Parties for the Elements of a Protocol or Another Legal Instrument, FCCC/AGBM/1997/2, January 31, 1997, at 43–44.

42. Synthesis of Proposals by Parties, Note by the Chairman, FCCC/AGBM/1996/10, November 19, 1996, at 9; Framework Compilation of Proposals from Parties for the Elements of a Protocol or Another Legal Instrument, Note by the Chairman, FCCC/AGBM/1997/2, January 31, 1997, at 35.

43. Additional Proposals from Parties, Note by the Secretariat, FCCC/AGBM/1997/Misc.1/Add.2, May 5, 1997, at 37.

44. See proposal by Iran, Framework Compilation of Proposals from Parties for the Elements of a Protocol or Another Legal Instrument, Note by the Chairman, FCCC/AGBM/1997/2, January 31, 1997, at 42; or comments by Venezuela and Iran at the fifth meeting of the AGBM, *Earth Negotiations Bulletin* 12:39, December 23, 1996, at 5.

45. Additional Proposals from Parties, Note by the Secretariat, FCCC/AGBM/1997/Misc.1/Add.2, May 5, 1997, at 24.

46. Additional Proposals from Parties, Note by the Secretariat, FCCC/AGBM/1997/Misc.1/Add.3, May 30, 1997, at 5.

47. Additional Proposals from Parties, Note by the Secretariat, FCCC/AGBM/1997/Misc.1/Add.3, May 30, 1997, at 1.

48. Proposals from Parties, Note by the Secretariat, FCCC/AGBM/1997/Misc.1/Add.10, November 12, 1997, at 3.

49. Proposals from Parties, Note by the Secretariat, FCCC/AGBM/1996/Misc.2/Add.2, November 15, 1996, at 28.

50. Proposals from Parties, Note by the Secretariat, FCCC/AGBM/1996/Misc.2/Add.2, November 15, 1996, at 30.

51. Proposals from Parties, Note by the Secretariat, FCCC/AGBM/1996/Misc.2/Add.2, November 15, 1996, at 31. Norway leaves the relative weighting of each factor open for further discussion.

52. Proposals from Parties, Note by the Secretariat, FCCC/AGBM/1996/Misc.2/Add.2, November 15, 1996, at 6.

53. Proposals from Parties, Note by the Secretariat, FCCC/AGBM/1996/Misc.2/Add.2, November 15, 1996, at 3.
54 Proposals from Parties, Note by the Secretariat, FCCC/AGBM/1996/Misc.2/Add.2, November 15, 1996, at 4.
55. For example, see Comments of China as reported in *Earth Negotiations Bulletin*, 12:97, November 16, 1998, at 3.

BIBLIOGRAPHY

Unpublished Material

American National Cattlemens Association. Papers. Entry 1713, American Heritage Center, University of Wyoming, Laramie.

Carpenter, Farrington R. Papers. Western History Collection, Denver Public Library.

Carpenter, Farrington R. *Transcript of Interview, May 21, 1955*. MSS#106, Colorado Historical Society, Denver.

Delaney, Frank. Papers. Norlin Library Archives, University of Colorado, Boulder.

Environmental Defense Fund. *Memos*, March 23 and 24, 1989 (copies on file with author).

Grounds, William. Papers. Entry 3024, American Heritage Center, University of Wyoming, Laramie.

U.S. Department of the Interior. *Office of the Secretary—Central Classified Files 1907–1936*. Record Group 48, Entry 749A, National Archives II, College Park, Maryland.

U.S. Department of the Interior. *Office of the Secretary—Central Classified Files 1907–1936*. Record Group 48, Entry 749B, National Archives II, College Park, Maryland.

U.S. Department of the Interior. *Office of the Secretary—Office Files of Secretary of the Interior Oscar Chapman 1933–1953*. Record Group 48, Entry 768, National Archives II, College Park, Maryland.

U.S. Department of the Interior. *Records of Interior Department Officials—General Records of First Assistant Secretary Theodore A. Walters, 1933–1937*. Record Group 48, Entry 777, National Archives II, College Park, Maryland.

U.S. Department of the Interior. *Records of Office of the Solicitor—General Correspondence and Other Records 1937–1958*. Record Group 48, Entry 810, National Archives II, College Park, Maryland.

U.S. Department of the Interior. *Records of Office of the Solicitor—Records of the Assistant Solicitor for Land Matters*. Record Group 48, Entry 812, National Archives II, College Park, Maryland.

U.S. Department of the Interior. *Records of Office of the Solicitor—Records of the Conservation Division Concerning the Grazing Service, 1934–1946*. Record Group 48, Entry 834, National Archives II, College Park, Maryland.

U.S. Department of the Interior. *Records of Office of the Solicitor—Records of the Conservation Division Concerning the Administration of the Taylor Grazing Act, 1935–1946*. Record Group 48, Entry 835, National Archives II, College Park, Maryland.

U.S. Department of the Interior. *Records of the Grazing Service—Central Files 1934–1946*. Record Group 49, Entry 3, National Archives, Washington, DC.

U.S. Department of the Interior. *Records of the Grazing Service—Correspondence Relating to Range Management 1934–1946*. Record Group 49, Entry 4, National Archives, Washington, DC.

U.S. Department of the Interior. *Records of the Grazing Service—Grazing License, Permit, and Appeals Case Files 1934–1946*. Record Group 49, Entry 6, National Archives, Washington, DC.

U.S. Department of the Interior. *Records of the Grazing Service—Public Relations Documents 1939–1942*. Record Group 49, Entry 33, National Archives, Denver, Colorado.

U.S. Department of the Interior. *Records of the Grazing Service—Miscellaneous Correspondence, 1934–1939*. Record Group 49, Entry 39, National Archives, Denver, Colorado.

U.S. Department of the Interior. *Records of the Grazing Service—Correspondence Relating to the Legislative Functions of the Division of Grazing 1934–1939*. Record Group 49, Entry 40, National Archives, Denver, Colorado.

U.S. Department of the Interior. *Records of the Grazing Service—Correspondence of F. R. Carpenter 1935–1939*. Record Group 49, Entry 41, National Archives, Denver, Colorado.

U.S. Department of the Interior. *Records of the Grazing Service—Correspondence Relating to the Organization of the Grazing Service, 1937–1944*. Record Group 49, Entry 42, National Archives, Denver, Colorado.

U.S. Department of the Interior. *Records of the Grazing Service—Correspondence of the Director's Office*. Record Group 49, Entry 44, National Archives, Denver, Colorado.

U.S. Department of the Interior. *Records of the Grazing Service—Range Management Subject File, 1930–1937*. Record Group 49, Entry 45, National Archives, Denver, Colorado.

U.S. Department of the Interior. *Records of the Grazing Service—Records of Hearings Concerning Grazing Privileges*. Record Group 49, Entry 270, National Archives, Denver, Colorado.

U.N. Framework Convention on Climate Change Secretariat. Official Documents, available on-line at http://maindb.unfccc.int/library/.

Wyoming Stock Growers Association. *Papers*. Entry 14, American Heritage Center, University of Wyoming, Laramie.

Wyoming Wool Growers Association. *Papers*. Entry 1350, American Heritage Center, University of Wyoming, Laramie.

Personal Interviews

Phil Barnett, staff member for Rep. Henry Waxman, November 19, 1999.

Dirk Forrister, former staff member for Rep. Jim Cooper, December 10, 1999.

Joseph Goffman, Environmental Defense Fund, October 1, 1999.

Judy Greenwald, former staff member for Rep. Phil Sharp, December 21, 1999.

Gary Hart, The Southern Co., November 3, 1999.

Ken Israels, EPA Region IX Office, July 1999.

Rusty Matthews, former aide to Sen. Robert Byrd, December 22, 1999.

Brian McLean, Director, EPA Acid Rain Division, October 25, 1999.

John McManus, American Electrical Power, January 17, 2000.

Sarah Wade, Environmental Defense Fund, September 27, 1999.

Allan Zabel, EPA Region IX Office, July 1999.

Government Publications

Buckman, Thomas E. 1937. Setting Up Taylor Grazing Districts in Nevada. Bulletin 77. University of Nevada Agricultural Extension Service. Entry 675-5, American Heritage Center, University of Wyoming, Laramie.

Congressional Research Service (CRS). 1994. *Market-Based Environmental Management: Issues in Implementation*, Report No. 94-213 ENR.

———. 1995. *Individual Transferable Quotas in Fishery Management*, by Eugene Buck, Report No. 95-849 ENR. September 25.

———. 2000. *The Law of the Sea Convention and U.S. Policy*, by Marjorie Ann Browne, Report No. IB95010. February 8.

Intergovernmental Panel on Climate Change (IPCC). 1990. *Climate Change: The IPCC Scientific Assessment*. Ed. J.T. Houghton, G.J. Jenkins, and J.J. Ephraums. Cambridge: Cambridge University Press.

———. 1996. *Climate Change 1995: Economic and Social Dimensions of Climate Change*. Ed. James P. Bruce, Hoesung Lee, and Erik F. Haites. Cambridge: Cambridge University Press.

———. 2001. *Climate Change 2001: The Scientific Basis*. Ed. J.T. Houghton, Y. Ding, D.J. Griggs, M. Noguer, P.J. van der Linden and D. Xiaosu. Cambridge: Cambridge University Press.

Organisation of Economic Co-operation and Development (OECD). 1992. *Climate Change: Designing a Tradeable Permit System*. Paris: OECD.

United Nations Conference on Trade and Development (UNCTD). 1992. *Combating Global Warming*. Geneva: UNCTD.

United Nations Framework Convention on Climate Change (UNFCCC). 1992. Available online at www.unfccc.int.

U.S. Congress. 1932. House, *Hearings on H.R. 11816*, Committee on the Public Lands, 72nd Congress, 1st Session. May 3, May 19, May 24, May 31, June 1, June 2.

———. 1933–1934. House, *Hearings on H.R. 6462*, Committee on the Public Lands, 73rd Congress, 2nd Session. June 7–9, 1933; February 19–21, February 23, February 28, March 1–3, 1934.

———. 1934. Senate, *Hearings on H.R. 6462*, Committee on Public Lands and Surveys, 73rd Congress, 2nd Session. April 20–May 2.

———. 1935. House, *Hearings on H.R. 3019*, Committee on the Public Lands, 74th Congress, 1st Session. March 1–12.

———. 1989. House, *Hearings on Clean Air Act Reauthorization (Part 2)*, Committee on Energy and Commerce, Subcommittee on Energy and Power, 101st Congress, 1st Session. September 12, October 4, October 11.

———. 1989. House, *Hearings on Clean Air Act Reauthorization (Part 1)*, Committee on Energy and Commerce, Subcommittee on Energy and Power, 101st Congress, 1st Session. September 7.

———. 1989. House, *Hearings on Clean Air Act Reauthorization (Part 3)*, Committee on Energy and Commerce, Subcommittee on Energy and Power, 101st Congress, 1st Session. October 18.

———. 1989. Senate, *Report on S. 1630*, Committee on Environment and Public Works, 101st Congress, 1st Session. December 20.

———. 1989. Senate, *Hearings on the Clean Air Act Amendments of 1989*, Environmental Protection Subcommittee, 101st Congress, 1st Session. September 26, October 3, October 4, October 9.

———. 1990. House, *Report on H.R. 3030*, Committee on Energy and Commerce, 101st Congress, 2nd Session. May 17.

———. 1990. Senate, *Hearings on the Energy Policy Implications of the Clean Air Act Amendments of 1989*, Senate Energy and Natural Resources Committee, 101st Congress, 2nd Session. January 24, January 25.

———. 1997. *Hearing before the Joint Economic Committee*, 105th Congress, 1st Session. July 9.

———. 1993. Senate, *A Legislative History of the Clean Air Act Amendments of 1990*, Committee on Environment and Public Works, 103rd Congress, 1st Session. November.

———. 1997. *Study on Tradable Emissions*, Joint Economic Committee, 105th Congress, 1st Session. July.

U.S. Department of the Interior. 1938. *Annual Report of the Secretary of the Interior.*

U.S. Environmental Protection Agency (U.S. EPA). 1998. *1997 Compliance Report: Acid Rain Program*, EPA-430-R-98-012.

———. 1999. *Draft Economic Incentive Program Guidance.* September.

———. 1999. *1998 Compliance Report: Acid Rain Division*, EPA-430-R-99-010.

———. 2000. Acid Rain Program Annual Progress Report, EPA-430-R-01-008..

U.S. General Accounting Office. 1994 (U.S. GAO). *Air Pollution: Allowance Trading Offers and Opportunity to Reduce Emissions at Less Cost*, GAO Report RCED-95-30.

———. 1997. *Air Pollution: Overview and Issues on Emissions Allowance Trading Program* (Testimony before Joint Economic Committee, Congress of U.S., July 7, 1997), GAO Report T-RCED-97-183.

U.S. National Acid Precipitation Assessment Program (NAPAP). 1991. *1990 Integrated Assessment Report.* Washington: NAPAP Office of the Director.

U.S. Office of Technology Assessment (OTA). 1985. *Acid Rain and Transported Air Pollutants: Implications for Public Policy.* Washington: Government Printing Office.

Watson, Robert T. 2000. Chair of the Intergovernmental Panel on Climate Change. *Presentation at the Sixth Conference of the Parties to the UNFCCC*, November 13.

NGO Publications

Earth Negotiations Bulletin. 1995–2000. Volume 12. Published by the International Institute for Sustainable Development (IISD).

Emissions Trading Education Initiative. 1999. *Real World Results.*

Environmental Defense Fund (EDF). 1997. *More Clean Air for the Buck: Lessons from the U.S. Acid Rain Emissions Trading Program*, by Daniel J. Dudek, Joseph Goffman, Deborah Salon, and Sarah Wade.

Environmental Law Institute (ELI). 1997. *Implementing an Emissions Cap and Allowance Trading System for Greenhouse Gases: Lessons from the Acid Rain Program.*

International Development Research Centre (IDRC). 1992. *For Earth's Sake: A Report from the Commission on Developing Countries and Global Change.* Ottowa: IDRC.

Pew Center on Global Climate Change. 1998a. *The Complex Elements of Global Fairness,* by Eileen Claussen and Lisa McNeilly.

Pew Center on Global Climate Change. 1998b. *Market Mechanisms and Global Climate Change,* by Annie Petsonk, Daniel J. Dudek, and Joseph Goffman.

World Resources Institute (WRI). 1998. *Contributions to Climate Change: Are Conventional Metrics Misleading the Debate?* by Duncan Austin, Jose Goldemberg, and Gwen Parker.

Newspapers and Magazines

The Boston Globe, September 19, 1989.

Denver Post, December 23, 1935.

The Economist, August 22, 1998.

High Country News, December 21, 1998.

Los Angeles Times, May 27, 1993.

The National Woolgrower, April 1932, August 1934, May 1935, December 1935, September 1936 (American Heritage Center, Entry 7889).

The New York Times, May 31, 2002; December 15, 1997; March 31, 1995; Aprril 8, 1993; March 14, 1993.

The Tribune, December 1, 1937 (American Heritage Center, Taylor Grazing Act Subject File #T212ga).

Wyoming Stockman-Farmer, Cheyenne, December 1929.

Unpublished Thesis

Kete, Nancy. 1992a. *The Politics of Markets: The Acid Rain Control Policy in the 1990 Clean Air Act Amendments.* Ph.D. dissertation, Johns Hopkins University.

Published Books and Articles

Ackerman, Bruce, and Richard Stewart. 1988. Reforming Environmental Law: The Democratic Case for Market Incentives. *Columbia Journal of Environmental Law* 13: 171.

Agrawal, Arun. 2002. Common Resources and Institutional Sustainability. In *The Drama of the Commons*, eds. Thomas Dietz, Nives Dolsak, Elinor Ostrom, and Paul C. Stern. Washington, DC: National Academy Press.

Agarwal, Anil, and Sunita Narain. 1991. *Global Warming in an Unequal World: A Case of Environmental Colonialism*. New Delhi: Center for Science and Environment.

Albin, Cecilia. 1995. Rethinking Justice and Fairness: The Case of Acid Rain Emission Reductions. *Review of International Studies* 21: 119.

———. 2000. *Justice and Fairness in International Negotiations*. Cambridge: Cambridge University Press.

Almond, Gabriel A. 1991. Rational Choice Theory and the Social Science. In *The Economic Approach to Politics*, ed. Kristen R. Monroe. New York: Harper Collins, 38–40.

Amy, Douglas J. 1984. Why Policy Analysis and Ethics Are Incompatible? *Journal of Policy Analysis and Management* 3(4): 573.

Anderson, Charles W. 1979. The Place of Principles in Policy Analysis. *The American Political Science Review* 73: 711.

Anderson, Elizabeth. 1993. *Value in Ethics and Economics*. Cambridge, MA: Harvard University Press.

Anderson, Terry L., and Peter J. Hill. 1975. The Evolution of Property Rights: A Study of the American West. *Journal of Law and Economics* 18(1): 163.

Anderson, Terry L., and Donald R. Leal. 2001. *Free Market Environmentalism*. New York: Palgrave Macmillan.

Andrews, Richard N.L. 1999. *Managing the Environment, Managing Ourselves: A History of American Environmental Policy*. New Haven, CN: Yale University Press.

Annala, John. 1996. New Zealand's ITQ System: Have the First Eight Years Been a Success or a Failure? *Reviews in Fish Biology and Fisheries* 6: 43-62.

Arrow, Kenneth, 1951. *Social Choice and Individual Values*. New York: Wiley.

Avineri, Shlomo. 1972. *Hegel's Theory of the Modern State*. Cambridge: Cambridge University Press.

Baer, Paul, et al. 2000. Equity and Greenhouse Gas Responsibility. *Science* 289: 2,287.

Barnes, Will. 1926. *The Story of the Range*. Washington, DC, Government Printing Office.

Baron, David P. 1990. Distributive Politics and the Persistence of Amtrak. *Journal of Politics* 52: 883.

Barrett, Scott. 1992. Acceptable Allocations of Tradeable Carbon Emissions Entitlements in a Global Warming Treaty. In *Combating Global Warming*. New York: UNCTD, 85–113.

Baslar, Kemal. 1998. *The Concept of the Common Heritage of Mankind in International Law*. The Hague: Martinus Nijhoff.

Bazerman, Max H. 1985. Norms of Distributive Justice in Interest Arbitration. *Industrial and Labor Relations Review* 38: 558.

Becker, Lawrence C. 1977. *Property Rights: Philosophic Foundations*. London: Routledge & Kegan Paul.

Bennett, Jane. 1987. *Unthinking Faith and Enlightenment: Nature and the State in a Post-Hegelian Era*. New York: New York University Press.

Bernstein, Mark, Alex Farrell, and James Winebrake. 1994. The Impact of Restricting the SO_2 Allowance Market. *Energy Policy* 22: 748.

Bodansky, Daniel. 1993. The United Nations Framework Convention on Climate Change: A Commentary. *Yale Journal of International Law* 18: 451.

Bohi, Douglas R., and Dallas Burtraw. 1997. SO_2 Allowance Trading: How Experience and Expectations Measure Up. Discussion Paper 97-24. Washington, DC: Resources for the Future.

Bricker, Owen P., and Karen C. Rice. 1993. Acid Rain. *Annual Review of Earth and Planetary Science* 21: 151.

Brod, Harry. 1992. *Hegel's Philosophy of Politics: Idealism, Identity, and Modernity*. Boulder, CO: Westview Press.

Bromley, Daniel W. 1991. Comment: Testing for Common Versus Private Property. *Journal of Environmental Economics and Management* 21: 92.

Brudner, Alan. 1995. *The Unity of the Common Law: Studies in Hegelian Jurisprudence*. Berkeley: University of California Press.

Bullard, Robert D. 1990. *Dumping in Dixie: Race, Class, and Environmental Quality*. Boulder, CO: Westview Press.

Burtraw, Dallas. 1998. Cost Savings, Market Performance, and Economic Benefits of the U.S. Acid Rain Program. Discussion Paper 98-28-REV. Washington, DC: Resources for the Future.

Burtraw, Dallas, and Byron Swift. 1996. A New Standard of Performance: An Analysis of the Clean Air Act's Acid Rain Program. *Environmental Law Review* 26: 10, 411.

Burtraw, Dallas, and Michael A. Toman. 1992. Equity and International Agreements for CO_2 Containment. *Journal of Energy Engineering* 118(2): 122.

Calef, Wesley. 1960. *Private Grazing and Public Lands*. Chicago: University of Chicago Press.

Carpenter, Farrington R. 1984. *Confessions of a Maverick*. Denver: Colorado Historical Society.

Carney, William J. 1998. From Stakeholders to Stockholders: A View from Organizational Theory. In *Who Owns the Environment?*, eds. Roger Meiners and P. J. Hill. Lanham, MD: Rowman & Littlefield, 187–221.

Chapman, Duane, and Thomas Drennen. 1990. Equity and Effectiveness of Possible CO_2 Treaty Proposals. *Contemporary Policy Issues* 8: 16–28.

Chinn, Lily N. 1999. Can the Market Be Fair and Efficient? An Environmental Justice Critique of Emissions Trading. *Ecology Law Quarterly* 26: 80.

Christman, John. 1994. *The Myth of Property: Toward an Egalitarian Theory of Ownership*. Oxford: Oxford University Press.

Coase, Ronald. 1960. The Problem of Social Cost. *Journal of Law and Economics* 3: 1-44.

Coggins, George Cameron, Charles F. Wilkinson, and John D. Leshy. 1993. *Federal Public Land and Resources Law.* 3rd ed. Westbury, NY: The Foundation Press.

Cohen, Morris. 1946. *The Faith of a Liberal*. New York: Henry Holt and Co.

———. 1949. *Studies in Philosophy and Science*. New York: Frederick Ungar Publishing.

———. 1959. *Reason and Nature: An Essay on the Meaning of Scientific Method.* Glencoe, IL: The Free Press.

———. 1967. *Law and the Social Order*. North Haven, CT: Archon Books.

Cohen, Richard E. 1995. *Washington at Work: Back Rooms and Clean Air*. Boston: Allyn and Bacon.

Coleman, James S. 1990. Norm-Generating Structures. In *The Limits of Rationality*, eds. Karen S. Cook and Margaret Levi. Chicago: University of Chicago Press.

Cook, Karen S., and Karen A. Hegtvedt. 1983. Distributive Justice, Equity, and Equality. *Annual Review of Sociology* 9: 217.

Craig, Paul P., Harold Glasser, and Willett Kempton. 1993. Ethics and Values in Environmental Policy: The Said and the UNCED. *Environmental Values* 2: 137.

Cramton, Peter, and Suzi Kerr. 1998. Tradable Carbon Permit Auctions: How and Why to Auction Not Grandfather. Discussion Paper 98-34 . Washington, DC: Resources For the Future.

Culhane, Paul J. 1981. *Public Lands Politics: Interest Group Influence on the Forest Service and the Bureau of Land Management*. Baltimore: Johns Hopkins University Press.

Dales, J.H. 1968. *Pollution, Property, and Prices*. Toronto: University of Toronto Press.

Dana, Samuel T., and Sally K. Fairfax. 1980. *Forest and Range Policy.* 2d ed. New York: McGraw-Hill.

Demsetz, Harold. 1967. Toward a Theory of Property Rights. *American Economic Review* 57: 347.

Dennen, R. Taylor. 1976. Cattlemen's Associations and Property Rights in Land in the American West. *Explorations in Economic History* 13: 423–36.

Downs, Anthony. 1957. *An Economic Theory of Democracy.* New York: Harper Press.

Drew, Elizabeth. 1978. *Senator*. New York: Simon and Schuster.

Dudek, Daniel J., and Alice LeBlanc. 1990. Offsetting New CO_2 Emissions: A Rational First Greenhouse Policy Step. *Contemporary Policy Issues* 8: 29.

Dudek, Daniel J., and John Palmisano. 1988. Emissions Trading: Why is this Thoroughbred Hobbled? *Columbia Journal of Environmental Law* 13: 217.

Dworkin, Ronald. 1977. *Taking Rights Seriously*. Cambridge, MA: Harvard University Press.

Dwyer John P., and Peter S. Menell. 1998. *Property Law and Policy: A Comparative Institutional Perspective*. Westbury, NY: The Foundation Press.

Eavey, Cheryl L. 1991. Patterns of Distribution in Spatial Games. *Rationality and Society* 3: 450.

Eavey, Cheryl L., and Gary J. Miller. 1984. Fairness in Majority Rule Games with a Core. *American Journal of Political Science* 28: 570.

Ellerman, A. Denny, Paul L. Joskow, Richard Schmalensee, Juan-Pablo Montero, and Elizabeth M. Bailey. 2000. *Markets for Clean Air: The U.S. Acid Rain Program*. Cambridge: Cambridge University Press.

Ellickson, Robert C. 1991. *Order without Law: How Neighbors Settle Disputes*. Cambridge, MA: Harvard University Press.

———. 1993. Property in Land. *Yale Law Journal* 102: 1,315.

Elster, Jon, ed. 1986. *Rational Choice*. Oxford: Basil Blackwell.

Elster, Jon. 1990. The Theory of Rational Choice. In *The Limits of Rationality*, eds. Karen Schweers Cook and Margaret Levi. Chicago: University of Chicago Press.

Epstein, Richard. 1998. Habitat Preservation: A Property Rights Perspective. In *Who Owns the Environment?*, eds. Roger Meiners and P.J. Hill. Lanham: Rowman & Littlefield.

Falen, Frank, and Karen Budd-Falen. 1993. The Right to Graze Livestock on the Federal Lands: The Historical Development of Western Grazing Rights. *Idaho Law Review* 30: 505.

Foss, Philip O. 1960. *Politics and Grass: The Administration of Grazing on the Public Domain*. Seattle: University of Washington Press.

Fullerton, Don, and Robert N. Stavins. 1998. How Economists See the Environment. *Nature* 395: 433.

Gardner, B. Delworth. 1984. The Role of Economic Analysis in Public Range Management. In *Developing Strategies for Rangeland Management*. Boulder, CO: Westview Press, 1,441–66.

Gates, Paul W. 1968. *History of Public Land Law Development*. Washington, DC: GPO.

Gillroy, John Martin. 2000. *Justice and Nature: Kantian Philosophy, Environmental Policy, and the Law*. Washington, DC: Georgetown University Press.

Goodin, Robert E. 1982. *Political Theory and Public Policy*. Chicago: University of Chicago Press.

———. 1994. Selling Environmental Indulgences. *Kyklos* 47: 573.

Graf, Michael. 1997. Application of Takings Law to the Regulation of Unpatented Mining Claims. *Ecology Law Quarterly* 24: 57–130.

Grafton, R. Quentin. 1996. Individual Transferable Quotas: Theory and Practice. *Reviews in Fish Biology and Fisheries* 6: 5–20.

Grant, Jill E. 1998. The Acid Rain Program. In *The Clean Air Act Handbook*, eds. Robert J. Martineau Jr. and David P. Novello. Chicago: American Bar Assoc..

Green, Donald P., and Ian Shapiro. 1994. *Pathologies of Rational Choice Theory*. New Haven: Yale University Press.

Grieder, William. 2001. The Right and U.S. Trade Law: Invalidating the 20th Century. *The Nation* (October 15).

Grubb, Michael, Christiaan Vrolijk, and Duncan Brack. 1999. *The Kyoto Protocol: A Guide and Assessment*. London: RIIA & Earthscan.

Grubb, Michael, and James K. Sebenius. 1992. Participation, Allocation and Adaptability in International Tradeable Emission Permit Systems for Greenhouse Gas Control, in *Climate Change: Designing a Tradeable Permit System*. Paris: OECD.

Hage, Wayne. 1989. *Storm over Rangelands: Private Rights in Federal Lands*. Bellevue, WA: Free Enterprise Press.

Hahn, Robert W. 1988. Innovative Approaches for Revising the Clean Air Act. *Natural Resource Journal* 28: 171.

———. 1989. Economic Prescriptions for Environmental Problems: How the Patient Followed the Doctor's Orders. *Journal of Economic Perspectives* 3: 95–114.

Hahn, Robert W., and Roger G. Noll. 1983. Barriers to Implementing Tradable Air Pollution Permits: Problems of Regulatory Interactions. *Yale Journal on Regulation* 1: 63.

Hampshire, Stuart. 1978. Morality and Pessimism. In *Public and Private Morality*, ed. Stuart Hampshire. Oxford: Oxford University Press, 1–22.

Hardin, Garrett. 1968. The Tragedy of the Commons. *Science* 162: 1,243.

Hartz, Louis. 1955. *The Liberal Tradition in America*. New York: Harcourt, Brace.

Hays, Sam. 1995. The Mythology of Emissions Trading. *Environmental Forum* 12: 14.

Hegel, G.W. 1967. *The Philosophy of Right*. Trans. T. M. Knox. Oxford: Oxford University Press.

Hochschild, Jennifer L. 1981. *What's Fair? American Beliefs about Distributive Justice*. Cambridge, MA: Harvard University Press.

Hockenstein, Jeremy B., Robert N. Stavins, and Bradley W. Whitehead. 1997. Crafting the Next Generation of Market-Based Environmental Tools. *Environment* 39: 12.

Hoffman, Elizabeth, and Matthew Spitzer. 1985. Entitlements, Rights, and Fairness: An Experimental Examination of Subjects' Concepts of Distributive Justice. *Journal of Legal Studies* 14: 259.

Hohfeld, Wesley. 1919. Fundamental Conceptions as Applied in Judicial Reasoning. *Yale Law Journal* 28: 721.

Horwitz, Morton. 1992. *The Transformation of American Law 1870–1960: The Crisis of Legal Orthodoxy*. Oxford: Oxford University Press.

Hume, David. 1978. *A Treatise of Human Nature*. 2d ed., Ed. L. A. Selby-Bigge and P. H. Nidditch. Oxford: Clarendon Press.

Jones, Tom, and Jan Corfee-Morlot. 1992. Climate Change: Designing a Tradeable Permit System. In *Climate Change: Designing a Tradeable Permit System*. Paris: OECD.

Joskow, Paul, and Richard Schmalensee. 1998. The Political Economy of Market-Based Environmental Policy: the U.S. Acid Rain Program. *Journal of Law and Economics* 41: 37.

Kalt, Joseph P., and Mark A. Zupan. 1984. Capture and Ideology in the Economic Theory of Politics, *American Economic Review* 74: 279.

Kau, James B., and Paul H. Rubin. 1993. Ideology, Voting, and Shirking. *Public Choice* 76: 151.

Kelman, Mark. 1988. On Democracy Bashing: A Skeptical Look at the Theoretical and 'Empirical' Practice of the Public Choice Movement. *Virginia Law Review* 74: 199.

Kelman, Steven. 1981. *What Price Incentives? Economists and the Environment*. Boston: Auburn House Publishing.

———. 1992. Cost-Benefit Analysis: An Ethical Critique. In *The Moral Dimensions of Public Policy Choice: Beyond the Market Paradigm*, eds. John Martin Gilroy and Maurice Wade. Pittsburgh: University of Pittsburgh Press, 153–64.

Keohane, Nathaniel, Richard L. Revesz, and Robert N. Stavins. 1998. The Choice of Regulatory Instruments in Environmental Policy. *Harvard Environmental Law Review* 22: 313–67.

Kete, Nancy. 1992b. The U.S. Acid Rain Control Allowance Trading System. In *Climate Change: Designing a Tradeable Permit System*. Paris: OECD.

Kinzig, Ann P., and Daniel M. Kammen. 1998. National Trajectories of Carbon Emissions: Analysis of Proposals to Foster the Transition of Low-Carbon Economies. *Global Environmental Change* 8: 183.

Klemme, Marvin. 1984. *Home Rule on the Range: The Early Days of the Grazing Service*. New York: Vantage Press.

Klyza, Christopher. 1996. *Who Controls Public Lands? Mining, Forestry, and Grazing Policies, 1870–1990*. Chapel Hill: University of North Carolina Press.

Kneese , Allen V., and Charles L. Schultze. 1975. *Pollution, Prices, and Public Policy*. Washington, DC: The Brookings Institution.

Leonard, Herman B., and Richard J. Zeckhauser. 1986. Cost-Benefit Analysis Applied to Risks: Its Philosophy and Legitimacy. In *Values at Risk*, ed. Duncan MacLean. Totowa, NJ: Rowman & Allanheld.

Liroff, Richard A. 1986. *Reforming Air Pollution Regulation: The Toil and Trouble of EPA's Bubble*. Washington, DC: Conservation Foundation.

Locke, John. 1960. *Two Treatises of Government*, ed. Peter Laslett. Cambridge: University of Cambridge Press.

Macauley, Molly K. 1997. Allocation of Orbit and Spectrum Resources for Regional Communications—What's at Stake? Discussion Paper 98-10. Washington, DC: Resources for the Future.

Macinko, Seth. 1993. Public or Private? United States Commercial Fisheries Management and the Public Trust Doctrine, Reciprocal Challenges. *Natural Resources Journal* 33: 919.

Macpherson, C.B. 1962. *Political Theory of Possessive Individualism*. Oxford: Clarendon Press.

———. 1977. Human Rights as Property Rights. *Dissent* 24: 72.

———. 1978. The Meaning of Property, in *Property: Mainstream and Critical Positions*, ed. C. B. Macpherson. Toronto: University of Toronto Press.

Mayer, Martin. 1966. *The Lawyers.* New York: Harper and Row.

McCay, Bonnie J., and James M. Acheson. 1987. *The Question of the Commons.* Tucson: University of Arizona.

McConnell, Grant. 1966. *Private Power and American Democracy.* New York: Alfred A. Knopf.

McKibben, Bill. 1989. *The End of Nature*. New York: Random House.

McLean, Brian J. 1997. Evolution of Marketable Permits: The U.S. Experience with Sulfur Dioxide Allowance Trading. *International Journal of Environment and Pollution* 8: 19.

Messick, David M., and Karen S. Cook. 1983. *Equity Theory: Psychological and Sociological Perspective.*

Merrill, Karen R. 2002. *Public Lands and Political Meaning: Ranchers, the Government, and the Property Between Them*. Berkeley: University of California Press.

Meyer, Aubrey. 1999. The Kyoto Protocol and the Emergence of 'Contraction and Convergence' as a Framework for an International Political Solution to Greenhouse Gas Emissions Abatement. In *Man-Made Climate Change—Economic Aspects and Policy Options*, eds. Olaf Hohmeyer and Klaus Rennings. Heidelberg: ZEW Economic Studies.

Mill, John S. 1978. Of Property. In *Property: Mainstream and Critical Positions*, ed. C. B. Macpherson. Toronto: University of Toronto Press.

———. 1998. *Utilitarianism*, ed. Quentin Crisp. Oxford: Oxford University Press

Miller, David. 1992. Distributive Justice: What the People Think. *Ethics* 102: 555.

Miller, Gary J. 1997. The Impact of Economics on Contemporary Political Science. *Journal of Economic Literature* 35: 1173.

Miller, Gary J., and Joe A. Oppenheimer. 1982. Universalism in Experimental Committee. *American Political Science Review* 76: 561.

Mohnen, Volker A. 1988. The Challenge of Acid Rain. *Scientific American* 259(2): 30.

Monroe, Kristen R. 1991. The Theory of Rational Action. In *The Economic Approach to Politics*, ed. Kristen R. Monroe. New York: Harper Collins.

Montgomery, David. 1972. Markets in Licenses and Efficient Pollution Control Programs. *Journal of Economic Theory* 5: 395.

Muhn, James, and Hanson R. Stuart. 1988. *Opportunity and Challenge: The Story of the BLM*. Washington, DC: U.S. Government Printing Office.

Munger, Michael C. 2000. *Analyzing Policy: Choices, Conflicts, and Practices*. New York: W. W. Norton.

Munzer, Stephen. 1990. *A Theory of Property*. Cambridge: Cambridge University Press.

Mwandosya, Mark J. 1999. *Survival Emissions: A Perspective From the South on Global Climate Change Negotiations*. Dar es Salaam: Center for Energy, Environment, Science, and Technology.

Nelson, Robert H. 1986. Private Rights to Government Actions: How Modern Property Rights Evolve. *University of Illinois Law Review* 2: 361–86.

Noll, Roger G., and Barry R. Weingast. 1991. Rational Actor Theory, Social Norms, and Policy Implementation, in *The Economic Approach to Politics*, ed. Kristen R. Monroe. New York: Harper Collins.

Nozick, Robert. 1974. *Anarchy State, and Utopia*. Cambridge, MA: Basic Books.

Nydegger, Rudy V., and Guillermo Owen. 1977. The Norm of Equity in a Three-Person Majority Game. *Behavioral Science* 22: 32.

Oberthur, Sebastian, and Herman E. Ott. 2000. *The Kyoto Protocol: International Climate Policy for the 21st Century*. Berlin: Springer Verlag.

Ostrom, Elinor. 1990. *Governing the Commons: The Evolution of Institutions for Collective Action*. Cambridge: Cambridge University Press.

Ott, Hermann E., and Wolfgang Sachs. 2000. *Ethical Aspects of Emissions Trading*. Wuppertal Papers No. 110.

Parson, Edward A., and Richard J. Zeckhauser. 1995. Equal Measures or Fair Burdens: Negotiating Environmental Treaties in an Unequal World. In *Shaping National Responses to Climate Change: A Post-Rio Guide*, ed. Henry Lee. Washington: Island Press.

Peffer, E. Louise. 1951. *The Closing of the Public Domain: Disposal and Reservation Policies 1900-50*. Stanford: Stanford University Press.

Peltzman, Sam. 1984. Constituent Interest and Congressional Voting. *Journal of Law and Economics* 27: 181.

Pigou, Arthur C. 1932. *The Economics of Welfare*. 4th ed. London: Macmillan and Co.

Porter, Gareth, and Janet W. Brown. 1996. *Global Environmental Politics*. 2d ed. Boulder: Westview Press.

Portney, Paul and John P. Weyant. 1999. *Discounting and Intergenerational Equity*. Washington, DC: Resources for the Future.

Posner, Richard. 1972. *Economic Analysis of Law*. Boston: Little, Brown.

Proudhon, Pierre-Joseph. 1994. *What Is Property?* Ed. Donald R. Kelley and Bonnie G. Smith. Cambridge, U.K.: Cambridge University Press.

Radin, Margaret Jane. 1982. Property and Personhood. *Stanford Law Review* 34: 957.

Rawls, John. 1972. *A Theory of Justice*. Oxford: Oxford University Press.

Raymond, Leigh. 1997. Viewpoint: Are Grazing Rights on Public Lands a Form of Private Property? *Journal of Range Management* 50(July): 431–38.

———. 2002. Localism in Environmental Policy: New Insights from an Old Case. *Policy Sciences* 35: 179.

Raymond, Leigh, and Sally K. Fairfax. 1999. Fragmentation of Public Domain Law and Policy: An Alternative to the "Shift-to-Retention" Thesis. *Natural Resources Journal* 39: 649.

Rees, Judith. 1990. *Natural Resources: Allocation, Economics, and Policy*. 2d ed. London: Routledge Press.

Reich, Charles. 1964. The New Property. *Yale Law Journal* 73: 733.

Riker, William H. 1982. *Liberalism against Populism: A Confrontation Between the Theory of Democracy and the Theory of Social Choice*. San Francisco: W. H. Freeman.

Roberts, Paul. 1963. *Hoofprints on Forest Ranges: The Early Years of National Forest Range Administration*. San Antonio: Naylor Co.

Rose, Adam. 1998. Viewpoint: Global Warming Policy: Who Decides What Is Fair? *Energy Policy* 26: 1.

Rose, Adam. 1990. Reducing Conflict in Global Warming Policy—The Potential of Equity as a Unifying Principle. *Energy Policy* 18: 927.

Rose, Adam, and Brandt Stevens. 1993. The Efficiency and Equity of Marketable Permits for CO_2 Emissions. *Resource and Energy Economics* 15: 117.

Rose, Carol M. 1994. *Property and Persuasion: Essays on the History, Theory, and Rhetoric of Ownership*. Boulder, CO: Westview Press.

Rose, Kenneth. 1997. Implementing an Emissions Trading Program in an Economically Regulated Industry: Lessons from the SO_2 Trading Program. In *Market-Based Approaches to Environmental Policy*, eds. Richard F. Kosobud and Jennifer M. Zimmerman. New York: Van Nostrand Reinhold, 101–124.

Rosenberg, Adam J. 1994. Emissions Credit Futures Contracts on the Chicago Board of Trade: Regional and Rational Challenges to the Right to Pollute. *Virginia Environmental Law Review* 13: 501.

Rowley, William D. 1985. *U.S. Forest Service Grazing and Rangelands: A History*. College Station, TX: Texas A&M University Press.

Rutstrom, E. Elisabet, and Melonie B. Williams. 2000. Entitlements and Fairness: An Experimental Study of Distributional Preferences. *Journal of Economic Behavior and Organization* 43: 75.

Ryan, Alan. 1984. *Property and Political Theory*. Oxford: Basil Blackwell.

———. 1987. *Property*. Minneapolis: University of Minnesota Press.

Sagar, Ambuj D. 2000. Wealth, Responsibility, and Equity: Exploring an Allocation Framework for Global GHG Emissions. *Climatic Change* 45: 511.

Sagoff, Mark. 1988. *The Economy of the Earth*. Cambridge: Cambridge University Press.

Sax, Joseph. 1983. Some Thoughts on the Decline of Private Property. *Washington Law Review* 58: 481.

Schelling, Thomas. 1960. *The Strategy of Conflict*. Cambridge, MA: Harvard University Press.

Schultze, Charles L. 1977. *The Public Use of Private Interest*. Washington, DC: Brookings Institution.

Sen, Amartya. 1992. *Inequality Reexamined*. Cambridge: Harvard University Press.

Shue, Henry. 1993. Subsistence Emissions and Luxury Emissions. *Law and Policy* 15: 39.

———. 1999. Global Environment and International Inequality. *International Affairs* 75(3): 531.

Smith, Kirk R. 1996. The Natural Debt: North and South. In *Climate Change: Developing Southern Hemisphere Perspectives*, eds. Thomas W. Giambelluca and Ann Henderson-Sellers. Chichester, U.K.: John Wiley and Sons, 423–48.

Solomon, Barry D. 1995. Global CO_2 Emissions Trading: Early Lessons from the U.S. Acid Rain Program. *Climate Change* 30: 75.

Solomon, Barry D., and Dilip R. Ahuja. 1991. International Reductions of Greenhouse-Gas Emissions: An Equitable and Efficient Approach. *Global Environmental Change* 1: 343.

Stavins, Robert N. 1997. *What Can We Learn from the Grand Policy Experiment? Positive and Normative Lessons from the SO_2 Allowance Trading*. Discussion Paper 97-45. Washington, DC: Resources for the Future.

Sterner, Thomas. 2002. *Policy Instruments for Environmental and Natural Resource Management*. Washington, DC: Resources for the Future.

Stigler, George J. 1971. The Theory of Economic Regulation. *Bell Journal of Economic and Management Science* 2: 3.

Stillman, Peter. 1980. Property, Freedom, and Individuality in Hegel's and Marx's Political Thought. In *NOMOS XXII Property*, eds. J. Roland Pennock et al. New York: New York University Press, 130–67.

Stokey, Edith, and Richard Zeckhauser. 1978. *A Primer for Policy Analysis*. New York: W. W. Norton.

Stone, Deborah. 2002. *Policy Paradox: The Art of Political Decision Making*. Rev. ed. New York: W. W. Norton.

Sunstein, Cass. 1993. Endogenous Preferences, Environmental Law. *Journal of Legal Studies* 22: 217–54.

Surazska, Wieslawa. 1986. Normative and Instrumental Functions of Equity Criteria in Individual Choices of the Input-Payoff Distribution Pattern. *Journal of Conflict Resolution* 30: 532.

Taylor, Prue. 1998. *An Ecological Approach to International Law: Responding to Challenges of Climate Change*. London: Routledge.

Thaler, Richard. 1991. Quasi-Rational Economics. New York: Russell Sage Foundation.

Tietenberg, Tom. 1985. *Emissions Trading: An Exercise in Reforming Pollution Policy*. Washington, DC: Resources for the Future.

———. 1988. *Environmental and Natural Resource Economics*. 2d ed. Glenview, IL: Scott, Foresman and Co.

———. 1992. Relevant Experience with Tradeable Entitlements. In *Combating Global Warming*. Geneva: UNCTD.

———. 2002. The Tradable Permits Approach to Protecting the Commons: What Have We Learned? In *The Drama of the Commons*, eds. Thomas Dietz, Nives Dolsak, Elinor Ostrom, and Paul C. Stern. Washington: National Academy Press.

Toth, Ferenc L., ed. 1999. *Fair Weather? Equity Concerns in Climate Change*. London: Earthscan.

Tribe, Lawrence H. 1972. Policy Science: Analysis or Ideology? *Philosophy and Public Affairs* 2: 66.

Victor, David G. 2001. *The Collapse of the Kyoto Protocol and the Struggle to Slow Global Warming*. Princeton: Princeton University Press.

Vogler, John. 2000. *The Global Commons: Environmental and Technological Governance*, second ed. New York: John Wiley and Sons.

Voigt, William. 1976. *Public Grazing Lands: Use and Misuse by Industry and Government*. New Brunswick, NJ: Rutgers University Press.

Waldron, Jeremy. 1994. The Advantages and Difficulties of the Humean Theory of Property. *Social Philosophy and Policy* 11: 85–123.

———. 1993. Property, Justification, and Need. *Canadian Journal of Law and Jurisprudence* 6(2): 185–215.

———. 1988. *The Right to Private Property*. Oxford: Clarendon Press.

———. 1986. John Rawls and the Social Minimum. *Journal of Applied Philosophy* 3: 21.

Walster, Elaine, and G. William Walster. 1975. Equity and Social Justice. *Journal of Social Issues* 31: 21.

Webb, Walter P. 1959. *The Great Plains*. Boston: Ginn and Company.

Weingast, Barry R. 1979. A Rational Choice Perspective on Congressional Norms. *American Journal of Political Science* 23: 245.

West, Patrick. 1982. *Natural Resource Bureaucracy and Rural Poverty: A Study in the Political Sociology of Natural Resources*. Ann Arbor, MI: University of Michigan School of Natural Resources Sociology Research Lab Monograph 2.

Williams, Bernard. 1998. A Critique of Utilitarianism. In *Moral Philosophy: A Reader*, 2d ed., ed. Louis Pojman. Indianapolis, IN: Hackett.

Williams, Howard. 1977. Kant's Concept of Property. *Philosophical Quarterly* 27: 32–40.

Young, H. Peyton. 1994. *Equity: In Theory and Practice*. Princeton: Princeton University Press.

Zafirovski, Milan. 2000. Extending the Rational Choice Model from the Economy to Society. *Economy and Society* 29:181.

Zajac, Edward E. 1978. *Fairness or Efficiency: An Introduction to Public Utility Pricing*. Cambridge, MA: Ballinger Publishing Co.

INDEX

ABOUT THE AUTHOR

Leigh Raymond is an Assistant Professor of Political Science at Purdue University. His research focuses on equity norms, property rights and localism in environmental policy. He has published articles in various scholarly journals, including: *Ecology Law Quarterly, The Journal of Range Management, Marine Policy, Natural Resources Journal, Policy Sciences,* and *Science.*